T0140818

Nonlinear Internal Model Control with Automotive Applications

Vom Promotionsausschuss der
Fakultät für Elektrotechnik und Informationstechnik
der
Ruhr-Universität Bochum
zur Erlangung des akademischen Grades
Doktor-Ingenieur
genehmigte Dissertation von

Dieter Schwarzmann

aus Temeschburg, Rumänien

Stuttgart, 2007

1st Reviewer: Prof. Dr.-Ing. Jan Lunze
Ruhr-Universität Bochum, Germany

2nd Reviewer: Prof.(i.R.) Dr.-Ing.Dr.h.c. Michael Zeitz
Universität Stuttgart, Germany

Thesis submitted on: 01.05.2007
Date of examination: 24.07.2007

Bibliografische Information der Deutschen Nationalbibliothek

Die Deutsche Nationalbibliothek verzeichnet diese Publikation in der Deutschen Nationalbibliografie; detaillierte bibliografische Daten sind im Internet über http://dnb.d-nb.de abrufbar.

ISBN 978-3-8325-1823-3

Logos Verlag Berlin GmbH
Comeniushof, Gubener Str. 47,
10243 Berlin
Tel.: +49 030 42 85 10 90
Fax: +49 030 42 85 10 92
INTERNET: http://www.logos-verlag.de

Acknowledgement

This dissertation is the result of four years of research as a student at the Ruhr-Universität Bochum in collaboration with Robert-Bosch GmbH.

The present achievement was made possible thanks to the excellent supervision of Prof. Lunze, who encouraged new ideas and approaches and constructively removed the bad ones. His work ethic, personal commitment and structured approach towards solving problems are amongst many values I chose to adopt.

A special thanks goes to Dr. Rainer Nitsche without whom I probably would have never started this Odyssee. His commitment and attitude towards advising PhD students allowed this research to be undisturbed by the daily routine encountered at the company, Robert Bosch GmbH. Of that company, I would like to thank Dr. Thomas Bleile and Dr. Martin Rauscher whose insight into function development and turbocharged engines was invaluable.

The largest contribution to the results of this thesis comes from all the students who have worked with me to solve the many problems. I am very fortunate, having had so many capable minds helping me. Especially the dedication of Andreas Schanz and Marco Schmidt will stay in my memory as we have spent insanely many nights in front of a white board trying to wrap our heads around the two-stage turbocharged air system. Many of the ideas presented in this thesis stem from one of those evenings and many more still need to be written down.

I would like to thank my colleagues at the Universität Bochum and my colleagues at Robert Bosch GmbH. I am indebted to Tobias Kreuzinger, Jochen Assfalg, Dr. Rainer Nitsche, Dr. Olivier Cois and Dr. Matthias Bitzer, who, together with Jonas Hanschke and Prof. Zeitz from the University of Stuttgart, sacrificed their time to allow me to bounce ideas off of them.

On a personal level, I would like to thank my parents and my friends for their patience and forgiveness during the time I committed myself to this work. I am deeply grateful for Lisa Nicole Henson, who accompanied me with her kindness and love through all the rough patches.

Thank you.

Stuttgart, January 2008 Dieter Schwarzmann

"So many new ideas are at first strange and horrible, though ultimately valuable that a very heavy responsibility rests upon those who would prevent their dissemination."

J. B. S. Haldane
(1892-1964, British geneticist and scientist)

"Eccentricity is not, as dull people would have us believe, a form of madness. It is often a kind of innocent pride, and the man of genius and the aristocrat are frequently regarded as eccentrics because genius and aristocrat are entirely unafraid of and uninfluenced by the opinions and vagaries of the crowd."

Dame Edith Sitwell
(1887-1964, English poet, critic and biographer;
Sister of Sir Osbert Sitwell)

"The engineer is the key figure in the material progress of the world. It is his engineering that makes a reality of the potential value of science by translating scientific knowledge into tools, resources, energy and labour to bring them into the service of man ... To make contributions of this kind the engineer requires the imagination to visualize the needs of society and to appreciate what is possible as well as the technological and broad social age understanding to bring his vision to reality. "

Sir Eric Ashby
(1904-1992, British botanist and administrator)

CONTENTS

Abstract

This work develops an internal model control (IMC) design method for nonlinear plants and employs this method to design pressure controllers for a one-stage and a two-stage turbocharged diesel engine.

The main focus lies on developing an applicable controller design method for automotive control problems. Automotive applications are characterised by a combination of the limited computational power of the car's on-board control unit and the nonlinear character of the systems to be controlled. Moreover, a required step in the development of series production controllers is the manual adaptation (calibration) of the controller parameters after the actual design. For this reason, the controller should provide tunable parameters. IMC is proposed as the control structure and the parameters of the internal model are chosen as tunable parameters.

The contribution of this thesis is two-fold. First, this work presents an IMC design procedure for nonlinear single-input, single-output systems. The nonlinear IMC, as proposed here, is based on the IMC structure known from linear systems and is based on a nonlinear feedforward control design. It is inversion-based and uses a low-pass state-variable filter which connects to the right inverse of the plant model to obtain a realisable IMC controller. Basic system properties, such as relative degree and internal dynamics, are exploited to extend the system class to stable and invertible plants. Input constraints and model singularities are taken into account by using a nonlinear low-pass filter that is made aware of the possible input/output behaviour of the model. This awareness is introduced by a model-dependent constraint of the filter's highest output derivative. The nonlinear IMC provides robust stability and robust tracking of the closed-loop system.

Second, the feasibility of this control scheme is presented. A single-input, single-output boost-pressure IMC controller is designed for a one-stage turbocharged diesel engine. The controlled plant was tested at the test bed and showed good results, surpassing the performance of the production PID-type controller. Two-stage turbocharging recently produced interest among car manufacturers and poses a challenging control problem due to the nonlinearity of the MIMO plant and a singularity of its inverse. This thesis presents the first model-based solution to this control problem. A multi-input, multi-output nonlinear IMC controller is designed and tested in simulations, showing good performance and robustness.

DEUTSCHE KURZFASSUNG

In der vorliegenden Arbeit wird eine Methode zum Entwurf eines nichtlinearen „Internal Model Control" (IMC) Reglers entwickelt. Diese Methode wird verwendet, um Ladedruckregler für einen einstufig sowie einen zweistufig aufgeladenen Motor zu entwickeln. In dieser Zusammenfassung wird beispielhaft auf die anwendung der IMC-Regelung bei dem zweistufig aufgeladenen Dieselmotor eingegangen.

Entwurf eines nichtlinearen SISO IMC-Reglers

IMC-Struktur

Abbildung 1 zeigt die IMC-Struktur mit IMC-Regler Q, Regelstrecke Σ und Modell der Regelstrecke $\widetilde{\Sigma}$. Die Arbeitsweise des IMC ist die Folgende: Ist das Modell $\widetilde{\Sigma}$ exakt ($\widetilde{\Sigma} = \Sigma$) und es treten keine Störungen auf ($d = 0$), so verschwindet das Rückführsignal ($y(t) - \tilde{y}(t) = 0$) und der IMC-Regler Q agiert als Vorsteuerung (s. Abb. 1b). Bei Modellfehlern und Störungen wirkt die Rückführung diesen Effekten automatisch entgegen. Für Systeme mit kleinen Modellunsicherheiten und Störungen ist es also wünschenswert, den IMC-Regler Q als Vorsteuerung zu entwerfen. Mit

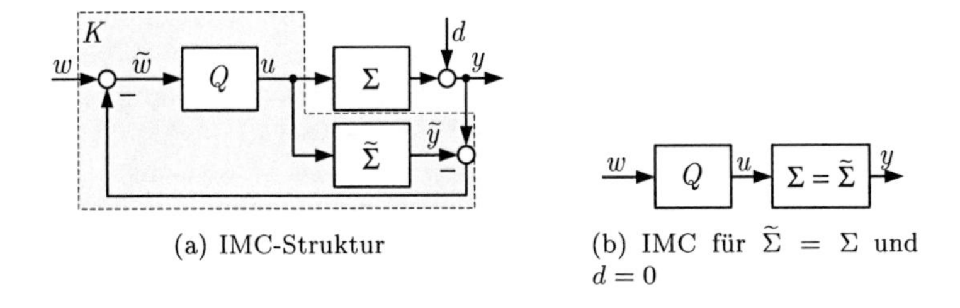

(a) IMC-Struktur

(b) IMC für $\widetilde{\Sigma} = \Sigma$ und $d = 0$

Abb. 1: IMC-Struktur.

$$\begin{aligned} \tilde{w} &= w - y + \tilde{y} \\ \tilde{y} &= \widetilde{\Sigma} \circ Q\, \tilde{w} \end{aligned} \tag{1}$$

findet man folgende strukturelle Eigenschaften der IMC-Struktur, welche auch für nichtlineare Systeme gelten [28]:

Eigenschaft 1 (Stabilität). *Wenn das Modell exakt ist ($\widetilde{\Sigma} = \Sigma$), ist der geschlossene Regelkreis in Abb. 1 stabil, wenn der IMC-Regler Q und die Regelstrecke Σ stabil sind.*

Eigenschaft 2 (Perfekte Regelung). *Der Ausgang $y(t)$ folgt der Führungsgröße perfekt ($y(t) = w(t)$, $\forall t > 0$) für beliebige Störungen $d(t)$, wenn der IMC-Regler Q die Inverse des Modells ist ($\widetilde{\Sigma} \circ Q = 1$) und der geschlossene Regelkreis in Abb. 1 stabil ist.*

Eigenschaft 3 (Verschwindende Regelabweichung). *Es entsteht keine bleibende Regelabweichung bei konstanten Führungsgrößen ($\lim_{t\to\infty} y(t) = \lim_{t\to\infty} w(t) = w$) und bei konstanten Störungen $\lim_{t\to\infty} d(t) = d$, wenn die stationäre Verstärkung Q_{ss} des IMC-Reglers invers zu der stationären Verstärkung des Modell ($\widetilde{\Sigma}_{\mathrm{ss}} \circ Q_{\mathrm{ss}} = 1$) und der Regelkreis in Abb. 1 stabil ist.*

Eigenschaften 2 und 3 gelten ohne explizite Forderung an die Modellgüte. Eigenschaft 3 besagt, dass eine verschwindende Regelabweichung erreicht wird, wenn die stationären Verstärkungen von Modell $\widetilde{\Sigma}$ und IMC-Regler Q invers zueinander sind. Es ist also nicht notwendig, einen expliziten Integralteil in den Regler einzuführen.

IMC-Entwurf für nichtlineare Systeme

Im Folgenden wird eine IMC Entwurfsmethode für nichtlineare Systeme vorgeschlagen. Hier wird auf Eingrößensysteme mit wohl definiertem relativen Grad und vorerst ohne Eingangsbeschränkungen eingegangen. Die nichtlineare Regelstrecke Σ, das nichtlineare Modell $\widetilde{\Sigma}$ der Regelstrecke und der IMC Regler Q werden als nichtlineare Operatoren betrachtet. Die Rechtsinverse $\widetilde{\Sigma}^{\mathrm{r}}$ des Modells $\widetilde{\Sigma}$ wird als Operator mit der Eigenschaft

$$\widetilde{\Sigma} \circ \widetilde{\Sigma}^{\mathrm{r}} \circ \tilde{y}(t) = \tilde{y}(t) \tag{2}$$

definiert. Die Rechtsinverse $\widetilde{\Sigma}^{\mathrm{r}}$ generiert die Eingangstrajektorie $u(t)$, so dass der Modellausgang $\tilde{y}(t)$ der gegebenen Trajektorie $\tilde{y}_{\mathrm{d}}(t)$ exakt folgt.

Ein nichtlinearer IMC-Regler Q kann nun angegeben werden als Verknüpfung eines *linearen* Tiefpassfilters F und der Rechtsinversen $\widetilde{\Sigma}^{\mathrm{r}}$

$$\boxed{Q = \widetilde{\Sigma}^{\mathrm{r}} \circ F.} \tag{3}$$

Bestimmung der Rechtsinversen. Die Rechtsinverse kann zum Beispiel (angelehnt an [35]) wie folgt bestimmt werden. Gegeben sei das Modell $\widetilde{\Sigma}$

$$\widetilde{\Sigma}: \quad \dot{x} = \boldsymbol{f}(\boldsymbol{x}, u), \quad \boldsymbol{x}(0) = \boldsymbol{x}_0 \tag{4a}$$

$$\tilde{y} = h(\boldsymbol{x}) \tag{4b}$$

mit relativem Grad r, welcher durch

$$r = \arg\min_r \left\{ \frac{\partial}{\partial u} L_f^r h(\boldsymbol{x}, u) \neq 0 \right\}$$

definiert ist. Hierbei stellt L_f die Lie-Ableitung entlang des Vektorfelds $\boldsymbol{f}$ dar. Mittels der Zustandstransformation

$$[y, \dot{\tilde{y}}, \ldots, \tilde{y}^{(r-1)}, \boldsymbol{\eta}]^T = \boldsymbol{\phi}(\boldsymbol{x}) \text{ mit} \tag{5a}$$

$$\tilde{y}^{(i)} = L_f^i h(\boldsymbol{x}) = \phi_{i+1}, \quad i = 0, \ldots, r-1 \tag{5b}$$

$$\boldsymbol{\eta} = \boldsymbol{\phi}_\eta(\boldsymbol{x}) \in \mathbb{R}^{n-r} \tag{5c}$$

kann das System $\widetilde{\Sigma}$ in die nichtlineare Ein-/Ausgangs-Normalform

$$\tilde{y}^{(r)} = \alpha\left(\tilde{y}, \dot{\tilde{y}}, \ldots, \tilde{y}^{(r-1)}, \boldsymbol{\eta}, u\right) \tag{6a}$$

$$\dot{\eta} = \boldsymbol{\beta}\left(\boldsymbol{\eta}, \tilde{y}, \dot{\tilde{y}}, \ldots, \tilde{y}^{(r-1)}, u\right) \tag{6b}$$

mit $\alpha(\cdot) = L_f^r h \circ \boldsymbol{\phi}^{-1}$ und $\beta(\cdot) = L_f \phi_{\eta,i} \circ \boldsymbol{\phi}^{-1}$ transformiert werden. Es wird angenommen, dass der relative Grad r wohl definiert ist und damit $\frac{\partial \alpha}{\partial u} \neq 0$ zumindest lokal gilt.

Aus der Ein-/Ausgangs-Normalform (6a) wird die Rechtsinverse $\widetilde{\Sigma}^{\mathrm{r}}$ als

$$\widetilde{\Sigma}^{\mathrm{r}}: \quad u = \alpha^{-1}\left(\tilde{y}_{\mathrm{d}}, \dot{\tilde{y}}_{\mathrm{d}}, \ldots, \tilde{y}_{\mathrm{d}}^{(r)}, \boldsymbol{\eta}\right) \tag{7a}$$

$$\dot{\eta} = \boldsymbol{\beta}\left(\boldsymbol{\eta}, \tilde{y}_{\mathrm{d}}, \dot{\tilde{y}}_{\mathrm{d}}, \ldots, \tilde{y}_{\mathrm{d}}^{(r-1)}, u\right) \tag{7b}$$

bestimmt.

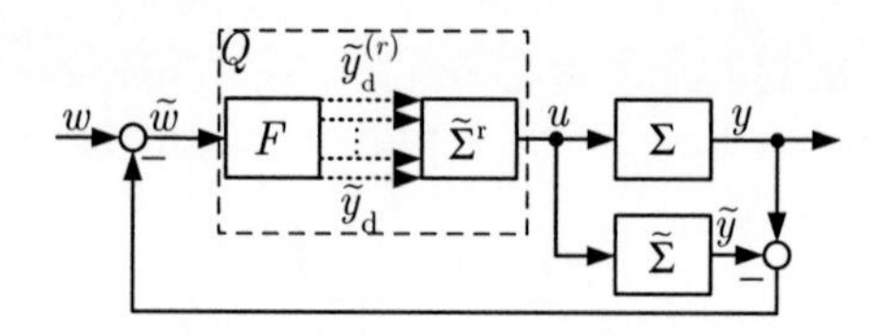

Abb. 2: IMC-Struktur nichtlinearer Systeme Σ.

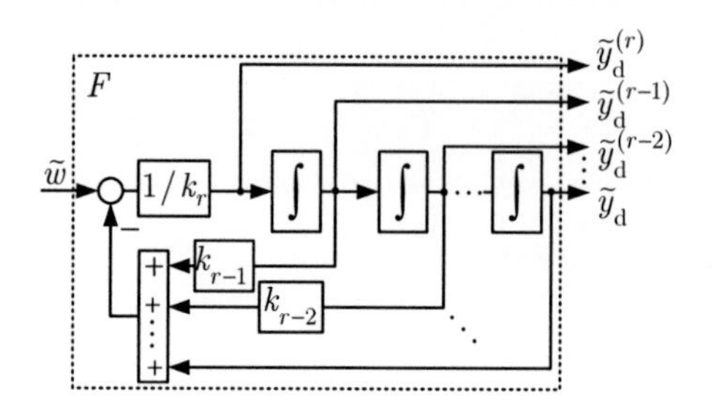

Abb. 3: IMC-Filter F.

IMC-Filter. Die Rechtsinverse $\widetilde{\Sigma}^{\mathrm{r}}$ benötigt den Wert der Solltrajektorie $\tilde{y}_{\mathrm{d}}$ sowie dessen erste r Zeitableitungen $\dot{\tilde{y}}_{\mathrm{d}}, \ldots, \tilde{y}_{\mathrm{d}}^{(r)}$. Diese müssen vom IMC-Filter F berechnet und der Rechtsinversen übergeben werden.

Die resultierende IMC-Struktur wird in Abb. 2 gezeigt. Damit das IMC-Filter F die Trajektorie $\tilde{y}_{\mathrm{d}}$ sowie die Information der Ableitungen $\dot{\tilde{y}}_{\mathrm{d}}$ bis $\tilde{y}_{\mathrm{d}}^{(r)}$ übergeben kann, wird vorgeschlagen, das IMC-Filter F als Zustandsvariablenfilter (siehe Abb. 3) mit der Übertragungsfunktion

$$F(s) = \frac{\tilde{y}_{\mathrm{d}}(s)}{\tilde{w}(s)} \frac{1}{k_r s^r + k_{r-1} s^{r-1} + \cdots + 1} \stackrel{!}{=} \frac{1}{(s/\lambda + 1)^r} \tag{8}$$

zu implementieren. Es ergibt sich somit als Differentialgleichung für den Filterausgang $\tilde{y}_{\mathrm{d}}$

$$k_r \tilde{y}_{\mathrm{d}}^{(r)} + k_{r-1} \tilde{y}_{\mathrm{d}}^{(r-1)} + \ldots + \tilde{y}_{\mathrm{d}} = \tilde{w}. \tag{9}$$

Zusammenfassung des IMC-Entwurfs. Der IMC-Entwurf für nichtlineare Systeme kann in den folgenden Schritten zusammengefasst werden:

1. Berechne die Rechtsinverse $\widetilde{\Sigma}^{\mathrm{r}}$ des Modells $\widetilde{\Sigma}$ der Regelstrecke Σ.

2. Entwerfe das IMC-Filter F wie in Gl. (8).

3. Implementiere das IMC-Filter als Zustandsvariablenfilter wie in Abb. 3.

4. Die resultierende Reglerstruktur ist in Abb. 2 dargestellt.

Der IMC-Regler Q für nichtlineare Systeme (4), welche in die Ein-/Ausgangs-Normalform transformiert werden können, setzt sich aus den Gleichungen (7) für die Rechtsinverse und der Gleichung (9) des Filters zusammen.

Die Existenz der Rechtsinversen ist eine notwendige Voraussetzung für den IMC-Entwurf. Die Berechnung der Rechtsinversen eines Modells ist abhängig von der Systemklasse des Modells. Die Bestimmung von Rechtsinversen für flache und Systeme in Ein-/Ausgangsnormalform werden in dieser Arbeit gezeigt.

Der dargestellte Entwurf eines nichtlinearen IMC ist eine direkte Erweiterung des linearen IMC-Entwurfs. Der resultierende Regelkreis basiert auf einer Ausgangsrückführung, ist nominell stabil und hat eine verschwindende bleibende Regelabweichung.

Berücksichtigen von Stellgrößenbeschränkungen. Sind die Stellgrößen $u \in [u_{\min}, u_{\max}]$ beschränkt, so kann das Filter geändert werden, dass es nur eine solche Wunschtrajektorie $\tilde{y}_{\mathrm{d}}$ produziert, welche mit der beschränkten Stellgröße vom Modell exakt erreicht werden kann ($\tilde{y}_{\mathrm{d}} = \tilde{y}$).

Die dafür notwendige Änderung erkennt man mittels Gl. (6a). Sind nämlich das Vektorfeld $\boldsymbol{f}$ sowie die Ausgangsfunktion h des Modells (4) analytische Funktionen von ihren Argumenten $\boldsymbol{x}$ und u, so folgt, dass die Funktion α auch analytisch (und damit stetig) in $\boldsymbol{x}$ und u ist. Somit bildet α bei gegebenen $\tilde{y}_{\mathrm{d}}, \ldots, \tilde{y}_{\mathrm{d}}^{(r-1)}$ das zusammenhängende Intervall der zulässigen Stellgröße $u \in [u_{\min}, u_{\max}]$ in ein zusammenhängendes Intervall für die zulässige höchste Ableitung $\tilde{y}_{\mathrm{d}} \in [\tilde{y}_{\mathrm{d,min}}^{(r-1)}, \tilde{y}_{\mathrm{d,max}}^{(r-1)}]$ ab. Dies geschieht mittels

$$\tilde{y}_{\mathrm{d,min}}^{(r)} = \min_{u \in [u_{\min}, u_{\max}]} \left\{ \alpha \left(\tilde{y}_{\mathrm{d}}, \dot{\tilde{y}}_{\mathrm{d}}, \ldots, \tilde{y}_{\mathrm{d}}^{(r-1)}, \boldsymbol{\eta}, u \right) \right\} \tag{10a}$$

$$\tilde{y}_{\mathrm{d,max}}^{(r)} = \max_{u \in [u_{\min}, u_{\max}]} \left\{ \alpha \left(\tilde{y}_{\mathrm{d}}, \dot{\tilde{y}}_{\mathrm{d}}, \ldots, \tilde{y}_{\mathrm{d}}^{(r-1)}, \boldsymbol{\eta}, u \right) \right\}. \tag{10b}$$

Wird das IMC-Filter wie in Abb. 4 verändert, so wird durch die Sättigungskennlinie gesichtert, dass nur solche Trajektorien $\tilde{y}_{\mathrm{d}}$ generiert werden, deren höchste Ableitungen $\tilde{y}_{\mathrm{d}}^{(r)}$ in dem erlaubten Invervall sind.

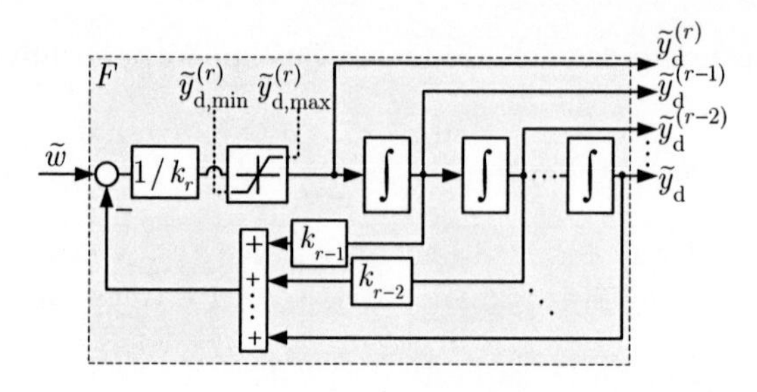

Abb. 4: Nichtlineares IMC-Filter mit Sättigungskennlinie zur Realisierung von Eingangsbeschränkungen.

Dadurch wird die Rechtsinverse nie eine Stellgröße u produzieren, welche die Stellgrößenbeschränkungen verletzt.

IMC Druckregelung eines zweistufig turboaufgeladenen Motors

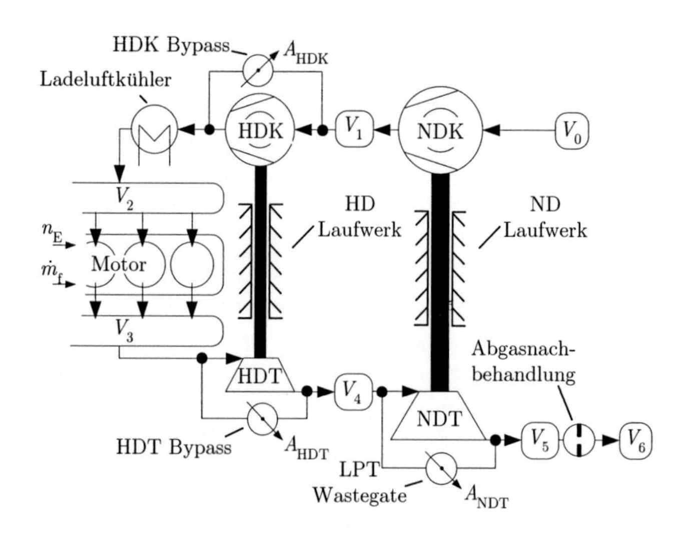

Abb. 5: Zweistufig aufgeladener Motor.

Ein zweistufig turboaufgeladener Motor mit Drehzahl n_{E} und Menge eingespritztem Kraftstoffs $\dot{m}_{\mathrm{f}}$ (siehe Abb. 5) hat zwei sequentiell angeordnete Turbolader. Jeder Turbolader wird als Stufe bezeichnet, wobei die Stufe am Motor die Hochdruckstufe (HD) und die Stufe an der Umgebung die Niederdruckstufe (ND) genannt wird. Die eingesaugte Luft wird zuerst vom Niederdruckkompressor (NDK) und dann vom Hochdruckkompressor (HDK) komprimiert. In derselben Weise wird der Abgasstrom zuerst über die Hochdruckturbine (HDT) und dann über die Niederdruckturbine (NDT) geleitet. Der Fluidstrom kann mittels Bypässen um den HDK als auch um beide Turbinen geleitet werden. Hierfür müssen die Öffnungsquerschnitte A_{HDK}, A_{HDT} und A_{NDT} entsprechend gestellt werden.

Es gilt, einen Regler zu finden, so dass die folgendenden Bedingungen erfüllt werden:

- Die gemessenen Drücke $\boldsymbol{y} = [p_2, p_3, p_4]^T$, welche in den Kammern V_1, V_2 und V_3 auftreten, folgen der Führungsgröße $\boldsymbol{w}$ ohne bleibende Regelabweichung.
- Die Stellgrößen sind die Querschnittsflächen der Bypässe $\boldsymbol{u} = [A_{\mathrm{HDK}}, A_{\mathrm{HDT}}, A_{\mathrm{NDT}}]^T$
- Die Stellgrößen $\boldsymbol{u}$ sind beschränkt durch $0 \leq u_i \leq u_{i\mathrm{max}}$ mit $i = 1, 2, 3$.

Die Eigenschaften der Regelstrecke erschweren die Regelungsaufgabe aus folgenden Gründen.

- Die Motordrehzahl n_{E} und die eingespritzte Kraftstoffmenge $\dot{m}_{\mathrm{f}}$ können für die Regelungsaufgabe als messbare Störungen interpretiert werden. Deren Wirkung muss vom Regler kompensiert werden.
- Es handelt es sich um eine Mehrgrößenregelung, so dass der zuvor vorgestellte IMC-Entwurf auf den Mehrgrößenfall angewendet werden muss.
- Die Regelstrecke verliert die Eigenschaft der Invertierbarkeit während des Betriebs, wenn der Ladedruck p_2 gleich dem Druck p_1 zwischen den Kompressoren ist, denn dann verliert der Eingang $u_1 = A_{\mathrm{HDK}}$ seine Wirkung auf das System.

Die Regelstrecke wird durch ein stabiles nichtlineares eingangsaffines Mehrgrößensystem repräsentiert. Der nichtlineare IMC nutzt die Ein-/Ausgangsnormalform für eingangsaffine Systeme, um die Rechtsinverse zu bestimmen. Die Stellgrößenbeschränkung wird mittels des limitierten

IMC-Filter F nach Abb. 3 realisiert, welches die Trajektorie $\tilde{y}_{\mathrm{d}}$ des Eingangs der Rechtsinversen $\widetilde{\Sigma}^{\mathrm{r}}$ so berechnet, dass diese ohne Verletzen der Stellgrößenbeschränkung erreicht werden kann. Die Singularität der Regelstrecke wird behandelt, indem der Regler das System genau so durch die Singularität steuert, dass diese Bewegung mittels vorhandener Stellgrößen erreicht werden kann. Dadurch ist die Rechtsinverse stets definiert.

Abbildung 6 zeigt das Simulationsergebnis des Ladedrucks p_2, des Abgasgegendrucks p_3 und des Drucks zwischen den Turbinen p_4 im Vergleich zu ihren Sollwerten w_1, w_2 und w_3. Ferner ist die Motordrehzahl n_{E} sowie der eingespritzte Kraftstoff $\dot{m}_{\mathrm{f}}$ als auch die aus dem IMC-Regler berechneten Bypassflächen u_1, u_2 und u_3 dargestellt. Die Simulation wurde so gewählt, dass sie realistische Betriebsbedingungen wiedergibt. Außerdem werden Modellfehler in der Simulation angenommen.

Die Simulationsergebnisse zeigen eine hohe Regelgüte als auch Robustheit gegenüber Modellfehlern.

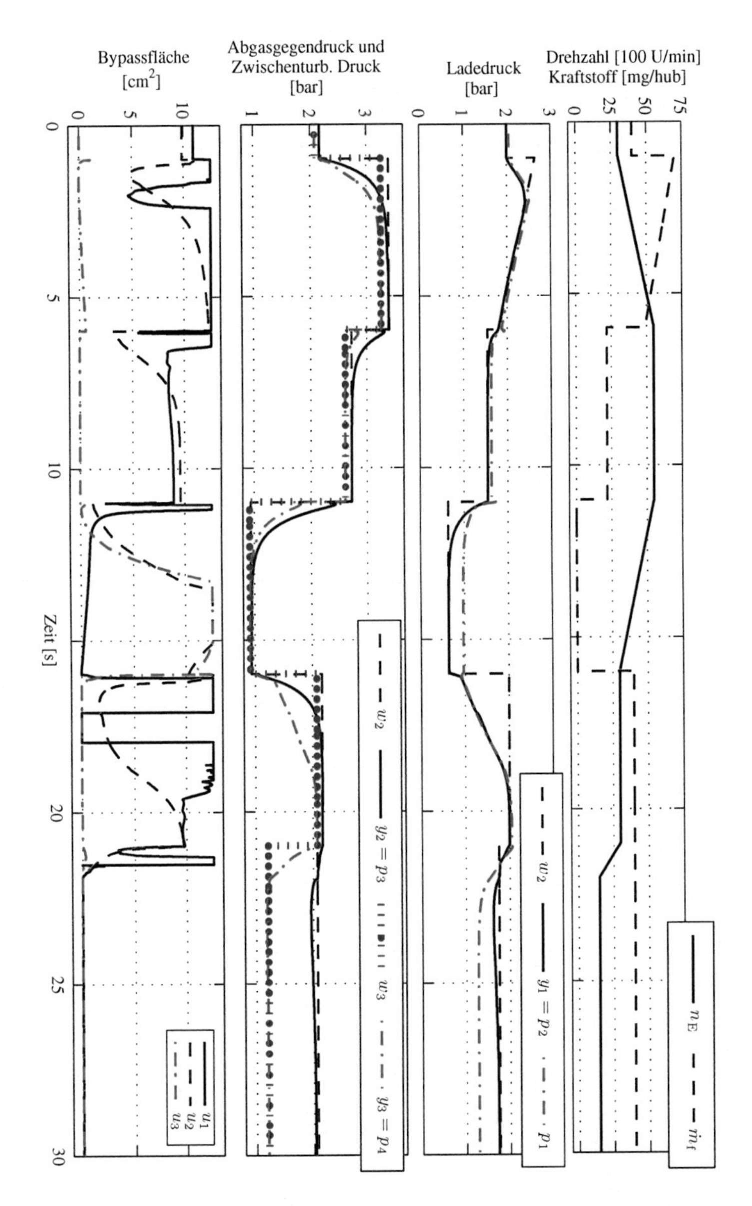

Abb. 6: Simulationsergebnis für einen realistisch parametrierten zweistufig aufgeladenen Dieselmotor mit IMC-Regelung.

NOMENCLATURE AND ABBREVIATIONS

A list of symbols is not given for the complete thesis, as some variables change their meaning in dependence on the context. All variables are properly labelled when used. An example is ended by the symbol ■, a definition is ended by the symbol ◇, and a proof is ended by □.

The commonly seen format for variables, vectors, and matrices are used (see e. g., [63] for a description). Additionally, the entries of a vector, say $\boldsymbol{x}$, are understood to be labelled as x_1, x_2, ..., and so forth. Thus, when introducing the variables x_1, x_2, ... it can be assumed, without an explicit introduction, that they belong to a vector $\boldsymbol{x}$.

Signals that are given in Laplace domain are denoted by small letters together with their dependency on the independent variable s, for example, $u(s)$ denotes the Laplace transform of the input signal $u(t)$. When it is desired to emphasise the linearity of a system, then this system is presented in its Laplace transform, for example, $G(s)$ means the linear and Laplace transformed system. The usual abuse of notation, namely not to use units in transfer functions or their poles and zeros, see e. g., [87], is employed. More precisely, within this text **time is understood to always have the unit of seconds and amplitude is dimensionless**, unless noted differently. Therewith, the independent variable s in Laplace domain has units of 1/seconds. However, in this text **transfer functions and their poles and zeros do not convey those units**.

Finally, the following table lists the frequently used abbreviations (acronyms) and their meaning.

List of abbreviations.

HPC	High-pressure compressor
I/O	Input-to-output, or input/output
IMC	Internal model control
ISE	Integral square error (norm)
LPC	Low-pressure compressor
MIMO	Multi-input multi-output
MP	Minimum phase
NMP	Non-minimum phase
OCU	On-board control unit
PID	Proportional Integral Derivative (Control)
RHP	Right-half (of the complex) plane
SISO	Single-input single-output
SVF	State-variable filter
VNT	Variable-nozzle turbine

1. INTRODUCTION

1.1 *Thesis Objective*

This thesis is motivated by control problems in the automotive industry. The objective is to develop a controller design method which is particularly suited for automotive control problems. Automotive control problems are characterised by a combination of the limited computational power of the car's on-board control unit and the nonlinear character of the systems to be controlled. The demands on closed-loop behaviour include performance criteria in time domain, such as rise time, settling time and overshoot. Moreover, a required step in the development of series production controllers is the manual adaptation (calibration) of the controller parameters. For this reason, the controller should provide tunable parameters.

The predominant controller design for automotive applications is a tedious process which relies heavily on trial-and-error. It is proposed to substitute this design method with a model-based controller design methodology by employing internal model control (IMC).

The classical IMC design concerns the control of linear systems. Although concepts to use IMC to control nonlinear systems exist, they do not share the IMC design philosophy and are not based on the IMC structure; as a consequence, they do not retain the typical IMC properties. This thesis proposes a nonlinear IMC design methodology which is based on the typical IMC design philosophy and employs the classical IMC structure. As a result of this, the nonlinear IMC inherits all closed-loop properties of the classical IMC, including robust stability and zero steady-state offset.

Two automotive control problems are solved using the proposed nonlinear IMC design method. They are the boost pressure control problem of a one-stage turbocharged diesel engine and a multi-input, multi-output pressure control problem of a two-stage turbocharged engine. In the case of the two-stage turbocharged engine, the proposed IMC presents the first published control solution to this problem which is fully able to utilise the behaviour of the plant.

1.2 Literature Review

Evaluation of relevant existing control design methods. The following briefly evaluates some controller design methods that are related to the IMC design as presented here. Therefore, some nonlinear control designs (often referred to as "constructive nonlinear control") [57, 83], such as construction of a Lyapunov function, backstepping, etc. are not discussed. Concerning the extensions proposed in Chapter 5, a topic-specific literature review is given when the individual extensions are introduced.

Internal model control. The internal model control structure is the basis for this thesis. It is presented in detail in Chapter 2. The main idea of IMC is that its structure degenerates to a feedforward control loop in the case of an exact model. The IMC design focuses on feedforward control design and relies on model inversion.

Concepts similar to IMC have been used since the late 1950s [50, 73] to design optimal feedback controllers. However, in 1974, it was first proposed [31] to use the typical IMC feedback structure to control processes. In 1982 [72], the IMC structure was employed in the field of robust control. IMC is at the core of the Smith Predictor [88] as well as some predictive control strategies [32]. It has been extended by an adaptive control strategy in [21] and is reviewed in some standard textbooks on control engineering like [63, 87]. Today, [72] is widely accepted as the standard reference of IMC. In 2002 [9], an IMC design procedure was introduced that guarantees robust performance of linear systems and presents an advanced linear IMC design method.

One finds that IMC meets all demands on an automotive controller, with the exception that its classical design method is limited to linear systems.

Nonlinear internal model control. The classical IMC design was extended to nonlinear systems (see e. g., [28, 31, 44, 101]). It was established already in the original work on IMC [31] in 1974 that the IMC structure is feasible to control nonlinear systems. The publications [28, 31] introduced a nonlinear IMC which relies on on-line numerical inversion of the model, based on iterative methods. Since this approach to nonlinear IMC uses a computational intense operation, it is likely that it cannot be executed in real-time by a car's on-board control unit (OCU).

In [44, 101], the goal is an exact I/O linearisation around which the IMC structure is placed. The contribution of the IMC structure to an I/O linearisation is that the internal model works as a feedforward state observer and that zero steady-state offset is guaranteed. However, this nonlinear IMC does not employ the typical IMC feedback structure, which does not have state feedback from the model. These nonlinear IMC approaches share the limitations of I/O linearisations as well as the limitations of IMC: The system class is limited to input affine, stable, and minimum phase[1] models. Thus, this approach to nonlinear IMC is unable to reproduce the classical IMC which, for example, is able to control non-minimum phase plants.

In summary, current extensions of IMC to nonlinear systems either use numerical inversion methods or do not employ the classical IMC structure or its design philosophy.

The main contribution of this work is the development of a general design concept for a nonlinear IMC which does not necessitate the feedback linearisation of the plant or plant model. It is based on some ideas presented in [28] as it also relies on the concept of the right inverse of the plant model. Using concepts from exact I/O linearisation and flat systems, the right inverse is obtained analytically and not numerically. As this inverse is, in general, not realisable since it depends on differentiations of its input signal, it is proposed to combine it with a low-pass filter. The composition of the filter with the inverse results in a non-anticipative system, which is realisable. The functional analytic interpretation of the closed-loop behaviour allows to extend the system class to which the nonlinear IMC is applicable. That is, it can treat input constraints, model singularities, and unstable model inverses. None of these can be treated by available approaches to nonlinear IMC. Moreover, the resulting control structure is identical to the classical IMC structure and can be designed and interpreted as the classical IMC.

The result is a nonlinear (IMC) controller which is robustly stable, guarantees zero steady-state offset, is straightforward to calibrate, respects input constraints, can handle model singularities, and is not restricted to minimum phase models.

[1] The property of (non-) minimum phase will, later on in this work, be defined as (in-)stability of the model inverse.

Exact Input-Output Linearisation Techniques The idea behind exact input-output (I/O) linearisation is to design a controller for a nonlinear plant such that the behaviour from the input of the controller to plant output is linear. The proposed nonlinear IMC uses some of the mathematical tools and concepts (in particular the notion of relative degree and the I/O normal form) that are used also in the context of I/O linearisation.

The majority of the literature on this topic (see e.g., [4, 5, 20, 41–43, 51, 54, 58, 60, 61]) focuses on input-affine nonlinear systems. An I/O linearisation uses a nonlinear model of the plant which is typically derived from physical laws and has physical parameters. Hence, a calibration of an I/O linearising controller consists of changing the physical model parameters which simplifies controller calibration. The computational burden is acceptable for an OCU since no on-line optimisation or similar computationally intense operations are needed. However, major drawbacks are the necessity of state feedback and the limitation to minimum phase plants. Moreover, robust stability of the closed-loop is difficult to prove.

If a system possesses the property of differential flatness (see e.g., [6, 29, 30, 39, 81]), it can elegantly be exploited for nonlinear controller design. The work of [39] addresses robustness issues of flatness-based controllers. The major advantage of flat systems is the simplicity with which a nonlinear feedforward control law can be designed [40, 102, 103].

In a two-degrees-of-freedom structure, a flatness-based feedforward controller can also be used with a linear PID feedback controller and, thus, also only necessitates output feedback.

Control of turbocharged engines.

One-stage turbocharged engine. There are numerous publications concerning the control of this plant. Work on control solutions of one-stage turbocharged engines includes [52], where a multi-input, multi-output linear parameter variant controller for both boost pressure and exhaust gas recirculation is presented. In [82], linear models for each operating point are computed through system identification based on test bed measurements. For each operating point, a linear model predictive controller is designed. The resulting controller is obtained by scheduling between the individual operating point-specific controllers. However, nominal stability is not guaranteed. In [71, 74] several linear control concepts for boost pressure and exhaust gas recirculation are introduced and evaluated. Of these, an H_∞ approach yields nominal and robust stability.

However, no dedicated calibration parameters are offered and one design iteration of an H_∞ controller can take several minutes on a standard PC and does not address time-domain behaviour directly; thus, calibration is time consuming. Some other solutions use nonlinear model predictive control [33, 45, 78, 95].

In conclusion, there is no current solution which offers calibration parameters that can be selected online at the test bed, is guaranteed to be nominally stable, and is inexpensive enough to be implemented in today's or future OCUs.

Two-stage turbocharged diesel engine. For the two-stage turbocharged diesel engine there exists no solution capable of using the full potential of the plant. However, there are two published approaches [90, 96] that deal with the control of a two-stage turbocharged engine. Both designs are essentially based on trial-and-error and control the MIMO plant of a two-stage turbocharged engine by a single-input, single-output switching controller which switches not only its parameterization but also its controlled variable. More detail on these publications is given in Section 7.3 since their assessment requires some introduction to the plant.

1.3 Controller Design in the Automotive Industry

1.3.1 Introduction

Today's automobiles are complex machines and perform a variety of tasks, beyond mere transportation. These tasks range from safety-related processes like anti-lock braking or deployment of an airbag; comfort-related functions like power steering, cruise control and air conditioning; and environmental-oriented functions like injection timing or catalytic conversion of exhaust gas. These tasks are controlled by programs (controllers) that run on the on-board control units (OCUs). Modern vehicles have nearly one hundred OCUs, each executing several control tasks.

Today, approximately (see [11])

- 90% of all innovations in automobiles are driven by electronics and software,
- 50%-70% of an OCU's development cost is determined by software, and

- 40% of a car's production cost stems from electronics and software development.

In addition, a car's OCUs collectively

- contain about *ten million* lines of code and
- run about 2500 features/tasks on
- four different bus systems concurrently.

From an engineering perspective, the automotive industry can be split into car suppliers and manufacturers. Most components of a car such as engine, power windows, drive train, electric motors, etc. are common to many cars and pose common control problems. The controllers are often developed by the supplier who delivers the respective OCU to several car manufacturers and therefore has to solve many control problems for components that share the same functionality (but not necessarily size or shape). Thus, the majority of control problems are not new but rather a variant of an older one. For such plants, a controller is obtained by manually adapting (calibrating) the available controller. As a result, today's controllers offer dedicated calibration parameters with which the closed-loop behaviour can be adjusted by a calibration engineer, with less effort than a complete new controller design. In order to do this efficiently, the function of the controller should be easy to comprehend. Today, there is an organisational separation between controller design and controller calibration. Hence, either the two are done by different departments in a single organisation, or they are done by different companies. Thus, if a modern control design method is proposed, it should still offer calibration parameters, since the process of calibration has become an important step in the development of the final product.

Since this thesis is concerned with methods for controller design, the requirements of the automotive industry must be considered. These requirements are discussed in detail in the following section.

1.3.2 Automotive Control Requirements

The goal of control design is to obtain a controller K which meets some predefined requirements on the closed-loop behaviour. Figure 1.1 shows a general control loop with controller K, plant Σ, control input u, disturbance d, reference signal w and plant output y. In this thesis, the plant Σ is assumed to be asymptotically stable. The closed-loop behaviour must

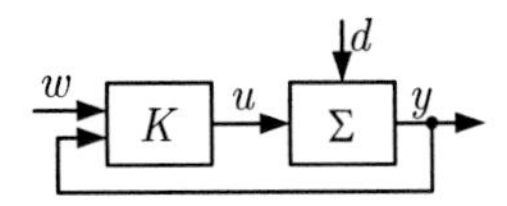

Fig. 1.1: General control-loop structure.

meet given specifications, despite some disturbance d that affects the plant in an undesirable way. Figure 1.2 shows the standard control loop with controller C, which can be used as a specific realisation of the general con-

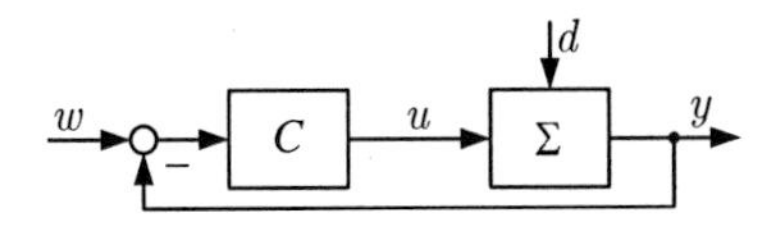

Fig. 1.2: Standard control loop.

trol loop shown in Fig. 1.1. The reference value w is influenced directly by the driver. Thus, future reference values of automotive control loops are unknown, which stands in contrast, for example, to the process industry in which the desired plant behaviour is defined by a given recipe.

In general, automotive control problems share the following properties and demands:

D1: Demands are posed in the time domain. Performance criteria concern properties like settling time, overshoot, and decay ratio of the closed-loop step response.

D2: The closed-loop behaviour should be robustly stable and obtain zero steady-state offset, despite part production tolerances and a constant disturbance.

The following are automotive-specific restrictions on the control law.

D3: The control algorithm has to be executed in real-time on the car's OCU[2].

D4: The controller must have adjustable parameters with which properties of the closed-loop system can be changed (calibrated), preferably

[2] Due to cost constraints, OCUs are computationally weak and offer little memory in comparison to a standard PC.

on-line, during a test bed run or during a test drive in an experimental vehicle. By manually calibrating the controller parameters, the controller should be able to control all structurally identical plants.

D5: Since parameters of the controller have to be calibrated, its functionality should be easily understood by the calibration engineer and the customer. Hence, the simplicity of a controller function is an important point to be considered.

Demands D1 and D2 are typically encountered in control engineering of many fields. Since the behaviour in time-domain is important, a control design, which can directly incorporate these demands in the design process, is preferred. A direct result of Demand D3 is that the controller should not rely on computationally intensive on-line optimisation procedures. Demands D4 and D5 are unusual in control engineering and stem from the unique requirements of the automotive industry.

Remark 1.1 (Additional requirements)**.** The implementation of a controller in an OCU requires two additional demands to be met. These are that the controller has to be represented by a sampling control algorithm and that, as of today, mainly fixed point arithmetic must be used.

Although both of these are important demands, they are not treated in this thesis explicitly. Concerning the necessity for a sampling control, it is assumed that the engineer is able to implement a continuous time controller in a sampling control algorithm. Since future OCUs will have the possibility to process floating point calculations, this requirement seemed less important and its treatment is assumed beyond the scope of this work.

The following chapter investigates and evaluates today's predominant controller design procedure.

1.3.3 Phenomenological Controller Design

The term "phenomenological" is used to qualify a process that relies mainly on repetitive observation and adaptation, in such that every change in the controller is motivated by an observed phenomenon of the plant. Thus, a phenomenological approach can be described as "trial-and-error".

Today's predominant controller design is done in a phenomenological process. This process consists of two phases. In the first phase, the control structure is chosen. In the second phase, the controller parameters are calibrated until the closed-loop behaviour satisfies the specifications.

Predominantly gain-scheduled PID controllers are used because of the clear implications of changes of the gains K_P, K_I, and K_D and the existence of tuning rules (e.g. Ziegler-Nichols) which provide closed-loop stability and some robustness (see [86] for a detailed investigation of the most popular tuning rules and [1] for an overview on gain scheduling). The transfer function $C(s)$ of a PID-controller, which is implemented in the standard control loop shown in Fig. 1.2, is given by $C(s) = K_P + \frac{K_I}{s} + \frac{K_D s}{1+T_1 s}$.

Control structure. Figure 1.3 shows the typical control structure. It

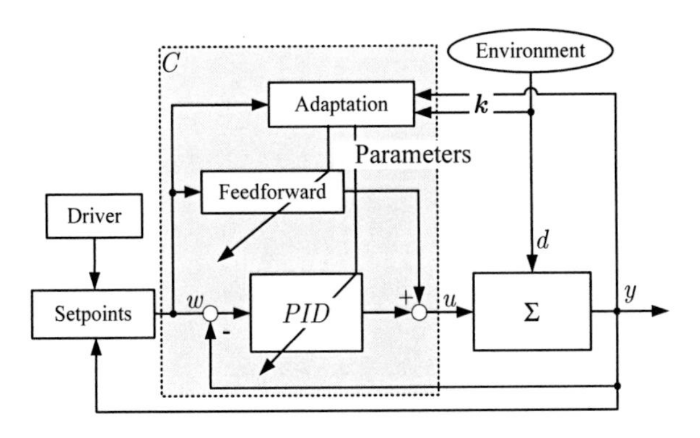

Fig. 1.3: Typical closed-loop control used to control automotive plants.

consists of a feedforward and a feedback path. However, the feedforward controller is not designed to improve performance but rather to compensate static nonlinearities: If the plant was linearised around the state $\boldsymbol{x}_{ss}$ and the input u_{ss}, the feedforward controller is a static block with reinforcement u_{ss}. Therefore, its degree-of-freedom is not used since it is determined by the steady-state behaviour of the plant Σ. Setpoints w are defined through the driver's demand and the current output y of the vehicle.

The gains, K_P, K_I, K_D are adapted in dependence upon the current operation mode and therein also in dependence upon the current operating point. The main idea, presented in the following in detail, is to partition the range of the measurement signals and then to find controller parameters in dependence of each partition. The following defines the notion of operation mode and operating point, as it is typically understood in the automotive industry, from a system theoretical perspective:

Operating point: Suppose that the chosen sensor signals are written in a vector $\boldsymbol{k} = [k_1, \ldots, k_p]^T$, where e. g., k_1 might be the current ambient temperature and k_2 could be the engine speed. Each entry k_i lies in a range $[k_{i\min}, k_{i\max}]$ with $i = 1, 2, \ldots, p$. This range is discretised into $m_i \in \mathbb{N}$ data points. Denote a discretised sensor signal as $\widehat{k}_i$. The current value of the vector of discretised sensor signals $\widehat{\boldsymbol{k}} = [\widehat{k}_1, \ldots, \widehat{k}_p]^T$ is the operating point. Note that the m_i data points of each discretised vector $\widehat{k}_i$ serve as nodes for look-up tables.

Operation mode: Define a set $\mathcal{K}$, here called the measurement space

$$\mathcal{K} = \{\boldsymbol{k},\ k_i \in [k_{i\min}, k_{i\max}]\}, \quad i = 1, 2, \ldots, p. \tag{1.1}$$

The set $\mathcal{K}$ consists of all possible combinations of measurement values. In order to define l operation modes, this set is divided into l distinct subsets $\mathcal{K}_\xi$ with $\xi = 1, \ldots, l$

$$\begin{aligned} \mathcal{K}_\xi \subset \mathcal{K},\ \text{where } \mathcal{K}_i \cap \mathcal{K}_j = \{\} \text{ for } i \neq j \\ \text{and } \bigcup_\xi \mathcal{K}_\xi = \mathcal{K}. \end{aligned} \tag{1.2}$$

When the current measurement lies in the ξ-th subset ($\boldsymbol{k}(t) \in \mathcal{K}_\xi$), one says that the plant is in operation mode ξ. Operation modes are used to vary the structure of the adaption algorithm. Figure 1.4 gives an example of two sensed signals k_1, k_2 which are divided to obtain five subsets $\mathcal{K}_1, \ldots, \mathcal{K}_5$, resembling five operation modes. The discretisation of $\widehat{k}_1$ and $\widehat{k}_2$ is indicated by the dotted lines.

The function of the adaptation algorithm can now be given as follows: Set the PID gains $K_\mathrm{P}, K_\mathrm{I}, K_\mathrm{D}$ by using an operation mode-specific function. For example, at the operation mode ξ one would write for the proportional gain K_P

$$K_\mathrm{P} = L_{\mathrm{P},\xi}(\boldsymbol{k}(t)), \tag{1.3}$$

where the value of the function $L_{\mathrm{P},\xi}$ is obtained from a look-up table in dependence upon the current measurement vector $\boldsymbol{k}(t)$. The nodes of the look-up table are given by the discretised vector $\widehat{\boldsymbol{k}}$.

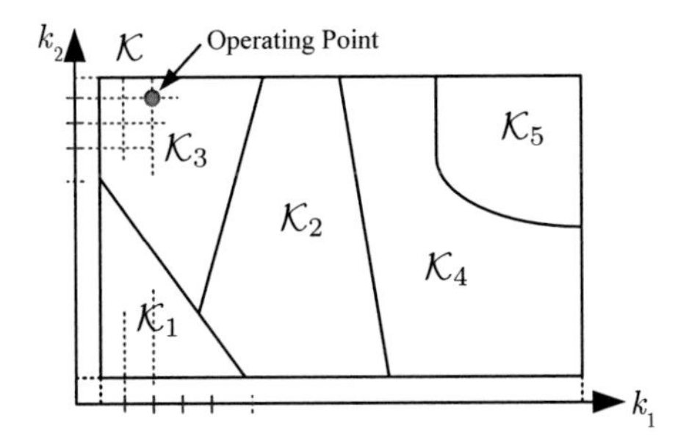

Fig. 1.4: Example of a measurement space $\mathcal{K}$ formed by $p = 2$ sensed signals divided into $\xi = 5$ subsets. The dotted lines indicate the discretisation of the measurement signals.

Hence, the design of the adaptation algorithm for a specific control problem includes the following steps:

1. Select an appropriate set of p available measurement signals k_i and their maximum and minimum values. This defines the measurement space $\mathcal{K}$.

2. Choose a discretisation $\widehat{k}_i$ of the measurement signals. This gives the nodes of the look-up tables. The intersection of the nodes gives the operating points.

3. Create an appropriate number of ξ operation modes and their boundaries.

4. Specify the structure of the adaptation functions $L_{\mathrm{P},\xi}, L_{\mathrm{I},\xi}, L_{\mathrm{D},\xi}$.

All of the above is done phenomenologically, i. e., by trial-and-error directly on a sample of the real plant. Thus, such a control design is a tedious process. From a system theoretical point of view, the aforementioned design method results in a switching controller, which is known [85] to not ensure performance, robustness or stability in neither the transition from one operating point to another nor in switching between operation modes.

Calibration. Before a controller can be employed on the final product, it needs to be calibrated. This means adjusting the scheduling tables' values (e.g. the parameters of the functions $L_{P,\xi}, L_{I,\xi}, L_{D,\xi}$), associated to each operating point, until a desired objective is met. Calibration is done manually in an iterative way for each entry in each table and for all operating points. Calibration is time-consuming and, therefore, an expensive procedure.

The current control structure fulfils the automotive-specific requirements D3-D5:

about D3: It uses few resources: The adaptation algorithm uses only basic arithmetic operations and the PID controller only has an order of two.

about D4: It provides calibration parameters: All values in the look-up tables can be adjusted.

about D5: It is easily comprehensible: The PID controller is quickly understood and the adaptation algorithm is developed using the experience gained by trial-and-error.

The following are *benefits* of the current development process.

- The exact behaviour of the plant at the test bed is considered, since the engineer develops the controller directly on the final product.
- Initial results are obtained quickly since, in each iteration, the controller is tested immediately on the real plant.
- The process is straightforward. The design of the adaptation algorithm does not require control engineering knowledge and calibration can be performed with basic understanding of the function of a PID controller.

Thus, this procedure seems to be an attractive method for automotive applications.

Some demerits of this control design method result from plant nonlinearities and from the heuristic design procedure. Both points contribute to the large adaptation complexity. Therefore, the *deficiencies* are the following:

- The plant complexity is not fully considered a priori by the control structure. Due to the phenomenological approach, even simple linear single-input, single-output plants (e. g., electronic throttle plates) are typically controlled using PID controllers with much more complex adaptation algorithms than necessary: Even if a single PID controller could control the plant, it is unlikely that this controller is found during the phenomenological design process.

- The major design effort lies in calibration. Several weeks are necessary to manually calibrate a controller for a moderately complex plant.
- Expertise, which is gained by individual engineers on a certain plant, cannot be preserved easily because it is intuitive, rather than instructive.
- There are no guarantees with respect to either stability or robustness. Although the final product goes through a series of tests on regular and extreme samples of the plant, failures are still encountered in the field.

In conclusion, today's phenomenological development procedure is becoming increasingly expensive for modern components. The following proposes a different approach to control design which promises to be more timely and, thus, less costly.

1.3.4 Model-Based Controller Design

The term "model-based control design" means that a (mathematical) model is used during development. However, it does not mean that the resulting controller includes this model. If a controller does include the model of the plant then it is called a "model-based controller".

Figure 1.5 shows the model-based control design procedure. It is assumed that an appropriate plant model already exists[3], and that the control problem is given. The control engineer then proceeds to select a control method (which may or may not result in a model-based controller) and a performance criterion with which this controller can be computed. The performance criterion is chosen to reflect the specifications of the closed- loop behaviour. The controller is then tested in simulations using the model of the plant. If it fails to pass the specifications, the performance criterion is altered. Once the controller fulfils the specifications, it can be tested at the test bed. The control design needs to provide

[3] The choice or development of a plant model is part of model-based control design. However, modelling is not a main topic of this thesis. For more information on modelling, the reader is referred to e. g., [53, 55] and the references therein.

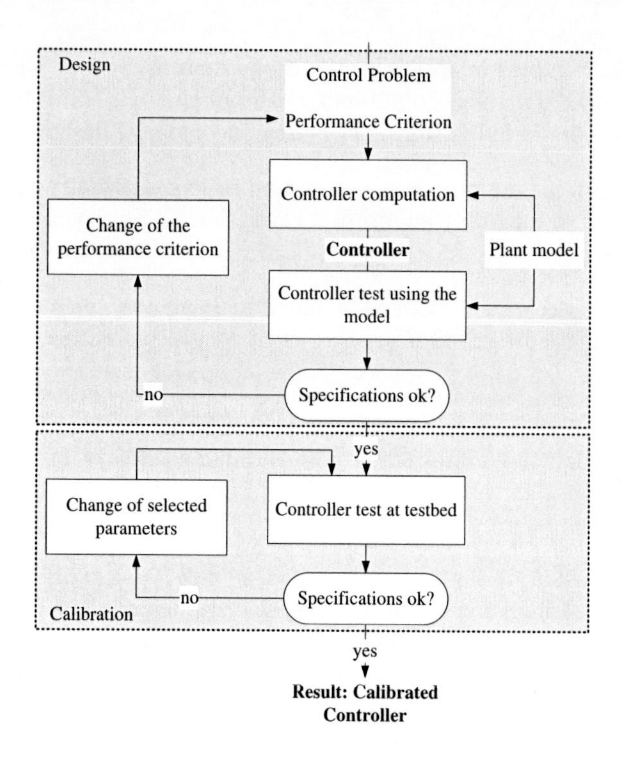

Fig. 1.5: Methodological controller development process.

some parameters with which the closed-loop behaviour can be calibrated. Thus, calibration is performed by adjusting only these parameters at the test bed.

The main advantage of the model-based design (Fig. 1.5) over the phenomenological design is a reduction in the number of iterations (trial-and-error) and a reduction in the number of calibration parameters. In the case of a good model and a good choice of the control method, there will be no iterations, whatsoever. The main advantages of a methodological design procedure are the following:

- The calibration procedure is reduced to few, possibly physically meaningful parameters, as well as distinct calibration parameters.
- Many influences (external and internal) do not have to be treated phenomenologically, since they are modelled and, thus, are automat-

ically treated by the controller.

- Guarantees about the resulting closed-loop behaviour and its robustness are possible for some control methods. Therefore, production tolerances for various plant components can be accounted for directly.

- The knowledge about the plant (i. e., its model) and the control concept is available explicitly and can be documented.

In conclusion, a methodological control design is well suited for automotive applications. However, not all control methods fulfil the demands D3-D5. In particular, a control method should be found to conclude the methodological control design process. This method should provide calibration parameters, is able to control nonlinear systems, addresses time-domain specifications directly, and does not use online-optimisations

1.4 Outline of the results.

This dissertation consists of two parts, preceded by this introduction.

Part I (Internal Model Control). Part I begins with a review of the classical IMC concept. The design of internal model control focuses on finding a controller that achieves a desired closed-loop input-to-output behaviour. It is based on feedforward control design and relies on model inversion. The IMC structure is simple and plausible and provides valuable properties such as nominal and robust stability as well as zero steady-state offset.

However, the classical IMC is limited to linear systems. The theoretical contribution of this thesis is the design methodology of a novel nonlinear IMC: This thesis proposes a nonlinear IMC controller to be designed as a feedforward controller by using the right inverse of the plant model together with a low-pass filter, called IMC filter. The connection of the two results in a realisable, non-predictive nonlinear feedforward controller. The requirement for this method is a stable and invertible plant model. In the IMC structure, this results in a nonlinear output feedback controller that is robustly stable and ensures zero steady-state offset. A functional-analytic view together with basic concepts of geometric control allows to extend the system class of the proposed nonlinear internal model control design method. The extensions include

- incorporating input constraints into the control design,
- guaranteeing finite inputs in the presence of singularities of the model inverse, and
- a method to control nonlinear systems with unstable inverses (non-minimum phase models).

The IMC design is shown in detail for flat systems and systems that can be transformed into the I/O normal form.

Part II (Internal Model Control of Turbocharged Engines). The model derivation of turbocharged engines as well as the reduction to control-oriented models are given in Appendix A. Part II exclusively deals with control design. The control-oriented model of the one-stage turbocharged air-system is flat, thus, a flatness-based IMC is developed. The IMC filter is chosen with respect to input constraints and the resulting controller was tested on a real engine at a test bed. The flatness-based IMC compares favourably to the current series production PID controller in terms of performance and calibration effort.

Finally, the IMC controller of a two-stage turbocharged air-system is developed. It provides tracking control of the boost pressure, the exhaust back pressure, and the pressure between the turbines of a two-stage turbocharged engine. Since the model is input-affine, the model inversion exploits the I/O normal form to develop the IMC controller. It is shown that invertibility of the plant model is lost if the pressure between the compressors equals the boost pressure. This implies a singularity of the model inverse. Due to the proposed singularity handling of the novel IMC, the controlled system never loses its relative degree: The setpoints are altered by the IMC filter automatically such that the resulting trajectories for the pressures can be achieved with the available inputs. A stability analysis shows that the closed-loop is robustly stable.

In conclusion, a novel nonlinear IMC controller is proposed and it is applied to two automotive control problems.

Part I

INTERNAL MODEL CONTROL

This part presents the theoretical contribution of the thesis. First, the control design method of internal model control (IMC) for linear SISO systems is reviewed and it is shown that it fulfils the demands of the automotive industry. Then, the IMC design method for nonlinear SISO systems is developed, with its specific application for flat and input-affine systems.

2. INTERNAL MODEL CONTROL OF LINEAR SISO SYSTEMS

The linear IMC (also referred to as classical IMC in this thesis) is introduced in detail in this chapter to establish a common basis for its extension to nonlinear systems. It is shown that IMC leads to a controller which meets all demands of the automotive industry.

2.1 *Structure and Properties of Internal Model Control (IMC)*

2.1.1 *Considered System Class*

In this chapter, the system class under consideration includes all linear time-invariant systems which can be described by transfer functions. All results also hold in the multi-input, multi-output (MIMO) case. However, to demonstrate the concept and design of IMC, *only* single-input, single-output (SISO) systems are regarded here. For a detailed discussion on the design of a MIMO IMC, the reader is referred to e. g., [72].

In the IMC design philosophy, plant Σ and plant model $\widetilde{\Sigma}$ are not considered equal. The IMC design uses only the model $\widetilde{\Sigma}$. Unlike most other control methods, IMC understands the plant Σ as part of the control loop but not as part of the control design. The plant Σ is interpreted as the real machine to be controlled and, as such, cannot be represented mathematically. The plant model $\widetilde{\Sigma}$ is used for control design. With such an interpretation, some model properties may be designed to intentionally differ from plant properties (see [9], where the model is not selected for plant match but for closed-loop robust performance). Therefore, the majority of this thesis is concerned with finding a controller based on the knowledge of the plant model $\widetilde{\Sigma}$.

The plant model $\widetilde{\Sigma}(s)$ is described by the transfer function

$$\widetilde{\Sigma}(s) = \frac{b_0 + b_1 s + \ldots + b_q s^q}{a_0 + a_1 s + a_2 s^2 + \ldots + a_n s^n}, \quad q \leq n \tag{2.1}$$

and consists of n poles p_i $(i = 1, \ldots, n)$ and q zeros z_j $(j = 1, \ldots, q)$.

Definition 2.1 (Relative degree of a transfer function, [63]). The relative degree r of a transfer function $\widetilde{\Sigma}(s)$ is the number of excessive poles

$$r = n - q. \tag{2.2}$$

◇

Systems with a relative degree equal to zero $(r = 0)$ are said to have a *direct feedthrough* (i. e., the system input u affects the system output y without delay).

2.1.2 *IMC Structure*

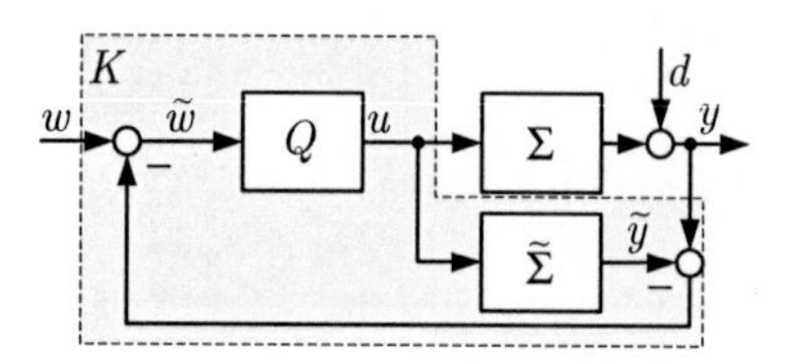

Fig. 2.1: IMC structure.

The IMC structure with *IMC controller* $Q(s)$ is shown in Fig. 2.1. The main idea of IMC is to include the model $\widetilde{\Sigma}(s)$ of the plant $\Sigma(s)$ into the controller $K(s)$, which qualifies IMC as a model-based controller. Note that the IMC structure is a specific realisation of the general control loop shown in Fig. 1.1 on page 7.

In the nominal case[1], one finds from Fig. 2.1 that the feedback signal vanishes $(y(t) - \tilde{y}(t) = 0, \ \forall t)$ and, consequentially, the IMC structure degenerates to a feedforward control where the IMC controller $Q(s)$ acts as a feedforward controller for the plant model $\widetilde{\Sigma}(s)$, as shown in Fig. 2.2. In the presence of a modelling error and the effect of a disturbance, the

[1] In the nominal case, there is no disturbance $(d = 0)$ and the plant model $\widetilde{\Sigma}(s)$ is an exact representation of the plant $(\Sigma(s) = \widetilde{\Sigma}(s))$.

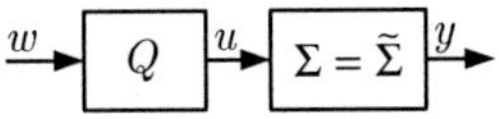

Fig. 2.2: Resulting IMC structure in the case of an exact model and no disturbance, i. e., the nominal case.

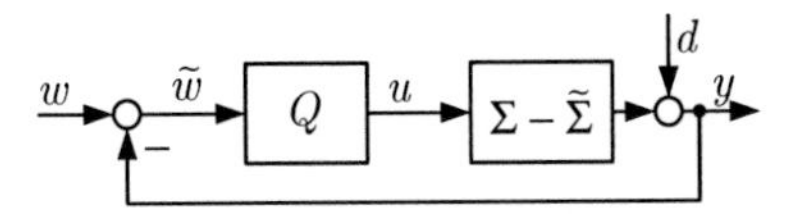

Fig. 2.3: Realisation of the IMC structure to visualise that Q acts on modelling errors $(\Sigma - \widetilde{\Sigma})$ and the disturbance d.

IMC structure from Fig. 2.1 is equal to the control structure shown in Fig. 2.3 which can be interpreted as a standard control loop (cf. Fig. 1.2) in which the feedback controller $Q(s)$ deals only with the modelling error $\Sigma(s) - \widetilde{\Sigma}(s)$ and the disturbance $d(t)$.

From Fig. 2.2 and 2.3, one concludes that an IMC controller $Q(s)$ essentially is a *feedforward controller* which is also used in feedback to attenuate the effect of the disturbance $d(t)$ and the modelling error $(\Sigma(s) - \widetilde{\Sigma}(s))$.

The IMC structure in Fig. 2.1 can also be implemented in a standard control loop (cf. Fig. 1.2), as shown in Fig. 2.4. One finds the follow-

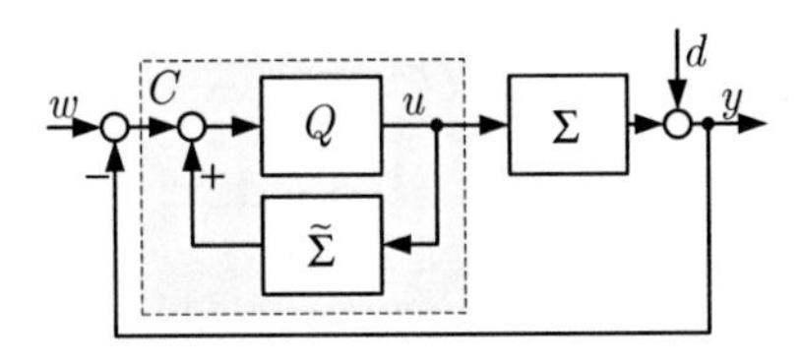

Fig. 2.4: IMC structure implemented in the standard control loop.

ing relationships between the IMC controller Q and a standard feedback controller C

$$C(s) = \left(1 - Q(s)\widetilde{\Sigma}(s)\right)^{-1} Q(s) \tag{2.3}$$

$$Q(s) = C(s)\left(1 + \widetilde{\Sigma}(s)C(s)\right)^{-1}. \tag{2.4}$$

Equations (2.3) and (2.4) show that every standard controller $C(s)$ can be realised by an IMC controller $Q(s)$ and vice versa. It is argued that an IMC controller $Q(s)$ can be designed in a straightforward fashion since the closed-loop specifications can be addressed directly. Moreover, it offers inherent structural closed-loop properties that are more difficult to obtain when directly designing a standard feedback controller $C(s)$.

2.1.3 *IMC Properties*

This section reviews general properties of IMC which result from the structure of the feedback loop shown in Fig. 2.1 and apply independently of the design method used to get the IMC controller $Q(s)$. They are independent in the sense that they only require $Q(s)$ to have basic properties, like stability and a unity steady-state gain.

The following three properties can be derived [72]:

Property 2.1 (Stability). *Assume the model to be exact ($\Sigma(s) = \widetilde{\Sigma}(s)$). Then, the closed-loop system in Fig. 2.1 is internally stable if and only if the controller $Q(s)$ and the plant $\Sigma(s)$ are stable.*

The proof closely follows that of [72] and is presented here for the sake of completeness.

Proof. In order to show internal stability, all transfer functions between the possible inputs and the possible outputs have to be stable. Figure 2.5

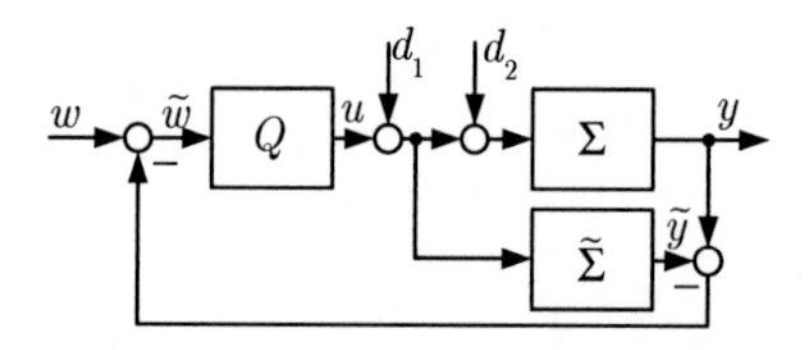

Fig. 2.5: IMC realisation used for internal stability conditions.

shows three independent inputs w, d_1, d_2 and one finds with $\widetilde{\Sigma}(s) = \Sigma(s)$ that

$$\begin{bmatrix} y(s) \\ u(s) \\ \tilde{y}(s) \end{bmatrix} = \begin{bmatrix} \Sigma(s)Q(s) & \Sigma(s) & (I - \Sigma(s)Q(s))\Sigma(s) \\ Q(s) & 0 & -\Sigma(s)Q(s) \\ \Sigma(s)Q(s) & \Sigma(s) & -\Sigma(s)^2Q(s) \end{bmatrix} \begin{bmatrix} w(s) \\ d_1(s) \\ d_2(s) \end{bmatrix} \tag{2.5}$$

holds. The transfer functions in Eq. (2.5) are stable if and only if both $Q(s)$ and $\Sigma(s)$ are stable. □

This result is to be expected, since the IMC structure degenerates to a feedforward control loop (Fig. 2.2) in the case of an exact model. An open loop control structure is only internally stable if each transfer function is stable. From this result, one concludes the following:

1. Only stable plants can be controlled with the IMC structure shown in Fig. 2.1: It is well known that an unstable plant can only be stabilised by feedback. Since the feedback signal of IMC may vanish, one would have to intentionally introduce modelling errors to establish a feedback signal which could then be used by a stabilising controller. Such a procedure, however, would defeat the purpose of a plant model.

2. Nominal stability of an IMC loop is trivially guaranteed for all stable plants $\Sigma(s)$ simply by chosing *any* stable transfer function $Q(s)$. This is a structural property of the IMC loop and it does not hold for a standard control loop (cf. Fig. 1.2) where stability of the closed-loop has to be be shown for each controller $C(s)$, even in the nominal case.

Property 2.2 (Perfect Control). *Assume that the IMC controller is equal to the* model *inverse ($Q(s) = \widetilde{\Sigma}^{-1}(s)$) and that the closed-loop system in Fig. 2.1 is stable. Then, the plant output $y(t)$ follows the reference signal $w(t)$ perfectly $y(t) = w(t)$, $\forall t$ for an arbitrary disturbance $d(t)$.*

Proof. An analysis of the block diagram in Fig. 2.1 gives

$$y(s) = \Sigma(s)\left(I + Q(s)\left(\Sigma(s) - \widetilde{\Sigma}(s)\right)\right)^{-1} Q(s)\cdot(w(s) - d(s)) + d(s), \tag{2.6}$$

which yields $y(s) = w(s)$ for $Q(s) = \widetilde{\Sigma}^{-1}(s)$ for arbitrary disturbances $d(s)$. □

The IMC property of perfect control holds in the presence of modelling errors and allows an interesting interpretation of the IMC structure: From the IMC loop in Fig. 2.1 one also finds

$$\frac{u(s)}{w(s)} = Q(s)\left(I + \Sigma(s)Q(s) - Q(s)\widetilde{\Sigma}(s)\right)^{-1}, \quad (d = 0), \tag{2.7}$$

which yields for $Q(s) = \widetilde{\Sigma}^{-1}(s)$ the closed-loop setpoint-to-input relationship

$$\frac{u(s)}{w(s)} = \Sigma^{-1}(s). \tag{2.8}$$

Thus, by chosing the IMC controller as the *model inverse* (and assuming closed-loop stability) the IMC structure essentially *inverts the plant behaviour* even though the plant behaviour is not exactly known.

However, perfect control cannot be realised in practise due to the following reasons:

1. Consider an IMC implemented in a standard control loop (cf. Fig. 2.4). One finds from Eq. (2.3) that if the IMC controller is a perfect inverse of the plant model ($Q(s) = \widetilde{\Sigma}^{-1}(s)$) the equivalent feedback controller $C(s)$ is mathematically not defined due to a division by zero.

2. If the model $\widetilde{\Sigma}(s)$ is non-minimum phase, its inverse is unstable and, thus, leads to an internally unstable closed-loop.

3. The model inverse is only realisable (proper) if the model has the same number of poles and zeros (i. e., if it has direct feedthrough). In such a case, however, the IMC structure would lead to a controller containing an algebraic loop which cannot be implemented, either.

Later on, the proposed IMC design will be interpreted as a low-frequency approximation of the perfect controller from above.

Property 2.3 (Zero Offset). *Assume that the steady-state controller gain is equal to the inverse of the steady-state model gain ($Q(0) = \widetilde{\Sigma}(0)^{-1}$) and that the closed-loop system in Fig. 2.1 is stable. Then, for any asymptotically constant reference signal $\lim_{t\to\infty} w(t) = w_{ss}$ and disturbance $\lim_{t\to\infty} d(t) = d_{ss}$, no steady-state control error occurs ($\lim_{t\to\infty} y(t) = w_{ss}$).*

Proof. Inserting $Q(0) = \widetilde{\Sigma}(0)^{-1}$ into Eq. (2.6) one finds $w(s) = y(s)$ for $s = 0$. □

Property 2.3 implies that if the steady-state gain of the IMC controller is the inverted steady-state gain of the model, the IMC structure has implicit integral action. Thus, it is not necessary (and rather unreasonable) to add an explicit integrator to the IMC controller to make the steady-state error vanish for constant $w(t)$ and $d(t)$. Unlike perfect control, which is not realisable, the property of zero steady-state offset is of practical use.

In the presence of model uncertainties, stability can be an issue. In order to show robust stability as a structural property, the Small-Gain Theorem [64, 99] is used. Consider a stable multiplicative output uncertainty[2] $\Delta(s)$ of the plant $\Sigma(s)$, as shown in Fig. 2.6. One finds the representation of the plant as

$$\Sigma(s) = (1 + \Delta(s))\widetilde{\Sigma}(s) \tag{2.9}$$

with the output uncertainty $\Delta(s)$. It is assumed that $\Delta(s)\widetilde{\Sigma}(s)$ is proper.

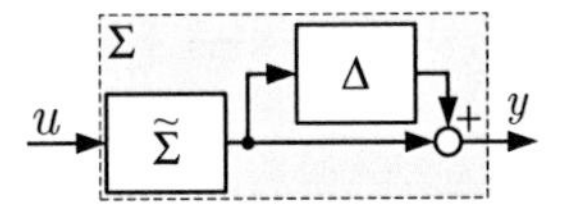

Fig. 2.6: Multiplicative output uncertainty Δ.

Property 2.4 (Robust Stability). *Assume that $\widetilde{\Sigma}(s)$ and $Q(s)$ are stable transfer functions, for which the IMC structure of Fig. 2.1 is stable for an exact model $\widetilde{\Sigma}(s) = \Sigma(s)$. Then, the IMC structure remains stable for multiplicative output uncertainties $\Delta(s)$ if*

$$\|\Delta(s)\widetilde{\Sigma}(s)Q(s)\|_\infty < 1 \tag{2.10}$$

holds, with $\|\cdot\|_\infty$ denoting the H_∞-norm[3].

Proof. The open-loop transfer function from the signal $\tilde{w}$ to the feedback signal $y - \tilde{y}$ is

$$\Sigma_0(s) \triangleq \frac{y(s) - \tilde{y}(s)}{\tilde{w}(s)} = \left(\Sigma(s) - \widetilde{\Sigma}(s)\right) Q(s). \tag{2.11}$$

[2] A multiplicative output uncertainty is a typical uncertainty description of the field of robust control. It is chosen here, since robust stability of an IMC loop can be shown in a straightforward fashion using the well-known Small-Gain Theorem.

[3] The H_∞-norm is defined as $\|G(s)\|_\infty = \sup_\omega \sigma_{\max}[G(j\omega)]$ where $\sigma_{\max}(\cdot)$ denotes the maximum singular value.

Inserting Eq. (2.9) into Eq. (2.11) results in

$$\Sigma_0(s) = \Delta(s)\widetilde{\Sigma}(s)Q(s).$$

The Small-Gain Theorem says that if the gain $|\Sigma_0(j\omega)|$ of the open-loop transfer function is smaller than one for all frequencies ω

$$\|\Sigma_0(s)\|_\infty < 1$$

then closing the loop yields a stable feedback system. □

Note that for a given upper boundary $\overline{\Delta}(\omega)$ of the amplitude $|\Delta(j\omega)|$ of the model uncertainty

$$|\Delta(j\omega)| \leq \overline{\Delta}(\omega), \quad \forall\, \omega, \tag{2.12}$$

the system is robustly stable if

$$\overline{\Delta}(\omega) < \frac{1}{|\widetilde{\Sigma}(j\omega)\, Q(j\omega)|}, \quad \forall\, \omega \tag{2.13}$$

holds.

In conclusion, the IMC controller $Q(s)$ works as a feedforward controller which is also used to attenuate modelling errors and disturbances. It has important properties like nominal and robust stability as well as offset-free control which makes it attractive for controller design. The following introduces a design method for internal model control for linear systems.

2.2 Classical IMC Design

The IMC design will be presented first for stable minimum phase (MP) systems. Then, it will be extended to non-minimum phase (NMP) systems.

The design presented here presents the standard design method as, for example, given in [87]. Note that this proposed design works well on many plants but still its performance can be improved by a more sophisticated approach [9, 72].

2.2.1 IMC Design of Minimum Phase Systems

Consider the model $\widetilde{\Sigma}(s)$ with relative degree r from Eq. (2.1) to be asymptotically stable and minimum phase. Thus, the n poles p_i $(i = 1, \ldots, n)$ and the q zeros z_j $(j = 1, \ldots, q)$ of the model $\widetilde{\Sigma}(s)$ are on the left half of the complex plane $(\mathrm{Re}\,(p_i)\,, \mathrm{Re}\,(z_j) < 0, \ \forall\, i, j)$.

Design. A feasible IMC controller $Q(s)$ for an NMP model $\widetilde{\Sigma}(s)$ with relative degree r is designed by

$$\boxed{Q(s) = \widetilde{\Sigma}^{-1}(s)F(s)}, \tag{2.14a}$$

with the *IMC filter* $F(s)$

$$F(s) = \frac{1}{\left(\frac{s}{\lambda}+1\right)^r} \quad \text{with } \lambda > 0. \tag{2.14b}$$

The IMC filter $F(s)$ has an r-fold pole at $-\lambda$. The variable λ is the only parameter to be chosen by the designer. Thus, with a given model $\widetilde{\Sigma}$, an IMC design focuses on choosing the parameter λ of the IMC filter. Here, it is proposed to place λ such that the desired closed-loop bandwidth[4] ω_B and high-frequency noise amplification coincides with that of the filter $F(s)$.

> **Remark 2.1.** Any stable transfer function $F(s)$ with the property $F(0) = 1$ and a relative degree of at least r would be feasible as an IMC filter. The presented IMC filter $F(s)$ is convenient since the value of only one parameter is to be chosen.

The resulting IMC structure is shown in Fig. 2.7.

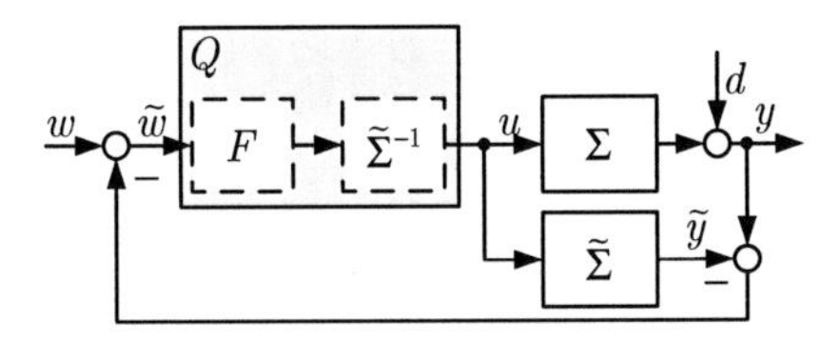

Fig. 2.7: IMC structure with IMC controller Q from Eq. (2.14a).

Explanation. An inverted model $\widetilde{\Sigma}^{-1}(s)$ is a perfect feedforward controller for $\widetilde{\Sigma}(s)$ and in the IMC structure, perfect control would follow (cf. IMC Property 2.2). However, since

$$\widetilde{\Sigma}^{-1}(s) = \frac{a_0 + a_1 s + a_2 s^2 + \ldots + a_n s^n}{b_0 + b_1 s + \ldots + b_q s^q} \tag{2.15}$$

[4] Let the bandwidth ω_B of a low-pass filter $F(s)$ with $|F(0)| = 0\text{dB}$ be defined as $|F(j\omega_\mathrm{B})| = -3\text{dB}$.

holds, the model inverse is not realisable if the model's relative degree is greater than zero ($r = n - q > 0$). In order to obtain a realisable transfer function $Q(s)$, the model inverse $\widetilde{\Sigma}^{-1}(s)$ is padded with the IMC filter $F(s)$, which compensates for the r excess zeros in Eq. (2.15) by introducing r additional poles. Those r additional poles are placed at the same location, namely at $-\lambda$, for convenience. The steady-state gain of the IMC filter is one ($F(0) = 1$) and it follows from Eq. (2.14a) that the steady-stage gains of the IMC controller is inverse to the steady-stage gain of the model (i. e., $Q(0) = \widetilde{\Sigma}^{-1}(0)$). Thus, an IMC loop with the IMC controller from Eq. (2.14a) will not have a steady-state offset (cf. IMC Property 2.3), even in the presence of model uncertainties.

Since the model $\widetilde{\Sigma}(s)$ is minimum phase, its inverse $\widetilde{\Sigma}^{-1}(s)$ is stable and nominal stability of the IMC loop with IMC controller $Q(s)$ from Eq. (2.14a) follows (cf. IMC Property 2.1). Considering robust stability, one finds with the IMC controller Q from Eq. (2.14a) and the stability criterion (2.10) with $\Sigma_0 = \Delta(s)F(s)$ that if

$$\|\Delta(s)F(s)\|_\infty < 1 \tag{2.16}$$

holds, then the closed loop is robustly stable with this IMC controller from which the condition

$$|F(j\omega)| < \frac{1}{\bar{\Delta}(\omega)}, \quad \forall\omega. \tag{2.17}$$

for robust stability can be derived. As the filter parameter can be chosen arbitrarily and $\|F(s)\|_\infty = \sup_\omega |F(j\omega)| = |F(0)| = 1$ holds, Eq. (2.17) shows a considerable robustness of the nonlinear IMC loop, because stability is ensured for $\sup(\bar{\Delta}(\omega)) < 1$.

Using the IMC controller $Q(s)$ from Eq. (2.14a) and assuming the plant to be represented using output uncertainties from Eq. (2.9), one finds with Eq. (2.6) that

$$\frac{y(s)}{w(s)} = T(s) = \frac{(I + \Delta(s))F(s)}{I - \Delta(s)F(s)} \tag{2.18}$$

holds with this controller. Thus, the closed-loop behaviour $T(s)$ tends to the behaviour of the IMC filter $F(s)$ for small model uncertainties $\Delta(s)$

$$\lim_{\Delta(s)\to 0} T(s) = F(s), \quad \forall s. \tag{2.19}$$

Interpretation. The IMC controller given in Eq. (2.14a) can be interpreted in several ways:

- The IMC controller $Q(s)$ is designed as feedforward controller for the model $\tilde{\Sigma}(s)$: A perfect feedforward controller is the inverse $\tilde{\Sigma}^{-1}(s)$ of the model. In order to be able to realise the IMC controller $Q(s)$, it is designed as an approximation of the model inverse $\tilde{\Sigma}^{-1}(s)$. The bandwidth ω_B of the IMC filter $F(s)$ governs which complex frequencies s of the model inverse $\tilde{\Sigma}^{-1}(s)$ are approximated well by $Q(s)$. Well below the IMC filter bandwidth the IMC controller $Q(s)$ given in Eq. (2.14a) is a good approximation of the model inverse $\tilde{\Sigma}^{-1}(s)$. This does not hold true for frequencies above the bandwidth of the IMC filter.

- The design of the IMC controller $Q(s)$ is equivalent to model reference control design: Model reference control [98] deals with finding a controller such that the closed-loop behaviour $T(s)$ is equal to a given stable reference model $R(s)$. Changing the IMC design in Eq. (2.14a) by setting the IMC filter $F(s)$ equal to the reference model $R(s)$

$$F(s) = R(s), \tag{2.20}$$

it follows from Eq. (2.19) that the presented IMC design is equivalent to model reference control for stable minimum phase plants.

Example 2.1 (IMC control of a linear minimum phase process):
Consider a plant model

$$\tilde{\Sigma}(s) = \frac{1}{s^2 + s + 1} \tag{2.21}$$

with no zeros and two complex conjugate poles at $p_{1/2} = -\frac{1}{2} \pm \frac{\sqrt{3}}{2} j$. From Eq. (2.14a) one finds the IMC controller

$$Q(s) = \underbrace{\frac{s^2 + s + 1}{1}}_{\tilde{\Sigma}^{-1}(s)} \underbrace{\frac{1}{\left(\frac{s}{\lambda} + 1\right)^2}}_{F(s)} \tag{2.22}$$

where $\lambda > 0$ is chosen by the designer to establish the desired closed-loop bandwidth. Now, consider a distorted plant

$$\Sigma(s) = \frac{10}{3s^2 + 5s + 7} \tag{2.23}$$

from which a multiplicative output uncertainty $\Delta(s)$ (cf. Eq. (2.9))

$$\begin{aligned}\Delta(s) &= \Sigma(s)\widetilde{\Sigma}^{-1}(s) - 1 \\ &= \frac{7s^2 + 5s + 3}{3s^2 + 5s + 7}\end{aligned} \tag{2.24}$$

can be derived. Figure 2.8 shows bode magnitude plots of the IMC filter $F(s)$ with the choice $\lambda = 1$, the uncertainty $\Delta(s)$ and the open loop transfer function $\Sigma_0 = \Delta(s)F(s)$. It shows that the closed IMC loop with this

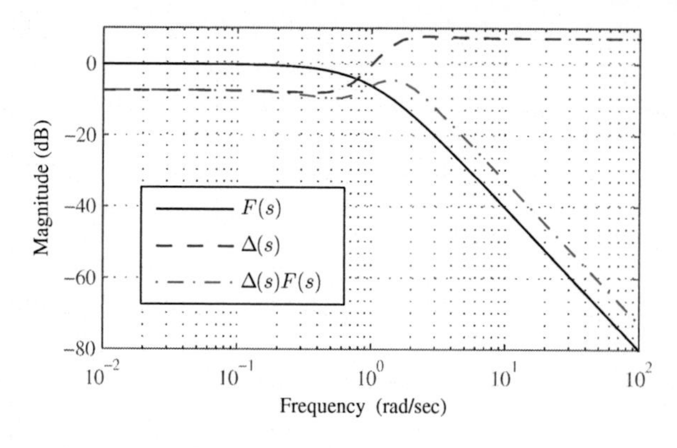

Fig. 2.8: Bode magnitude plot of IMC filter $F(s)$, uncertainty $\Delta(s)$ and open loop $\Sigma_0(s) = \Delta(s)F(s)$, given in Example 2.1.

uncertainty $\Delta(s)$ and IMC filter $F(s)$ is stable since $\|\Delta(s)F(s)\|_\infty < 0\text{dB} = 1$ holds. The high frequency magnitude of the uncertainty $\Delta(s)$ is above one ($\lim_{\omega\to\infty}|\Delta(j\omega)| = \frac{7}{3}$), but the IMC filter attenuates these high frequency gains. Thus, the choice of the IMC filter pole λ determines not only the closed-loop bandwidth but also robust stability (see [63, 72] for an in-depth discussion on this matter). Finally, Fig. 2.9 shows the step response $y_\text{d}(t)$ of the nominal case and the case with the modelling errors $y(t)$ assuming the plant from Eq. (2.24). It is assumed that the shape of both step responses are acceptable since they are non-oscillatory. As expected, zero steady-state offset is achieved even for the distorted plant. ■

2.2.2 IMC Design of Non-Minimum Phase Systems

Consider the model $\widetilde{\Sigma}(s)$ with relative degree r from Eq. (2.1) to be stable and NMP. Thus, it either has a time delay or zeros in the right half plane. Time delays are not considered in this work. The basic idea of dealing with an NMP system is the following:

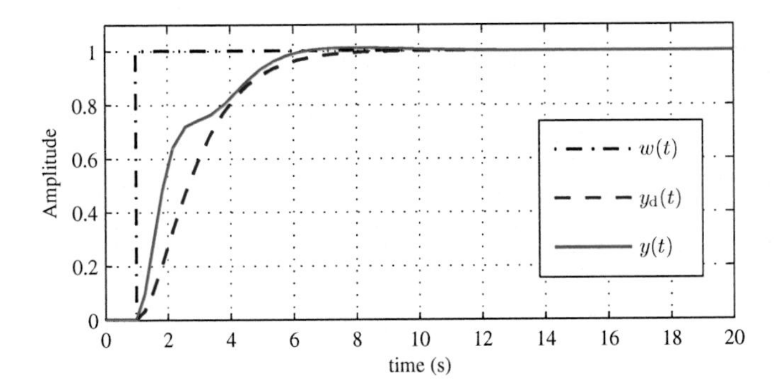

Fig. 2.9: Response of the nominal output $\tilde{y}_\mathrm{d}$ and actual output y of Example 2.1, to a reference step in w.

1. Split the model $\widetilde{\Sigma}(s)$ into two parts where one part $\widetilde{\Sigma}_\mathrm{MP}(s)$ is MP and the other part $\widetilde{\Sigma}_\mathrm{NMP}(s)$ contains the NMP behaviour.

2. Design the IMC controller $Q(s)$ as described in Eq. (2.14a) but only for the MP part $\widetilde{\Sigma}_\mathrm{MP}(s)$.

3. Implement the original NMP model $\widetilde{\Sigma}(s)$ as internal model in the IMC structure.

By inverting only the MP part $\widetilde{\Sigma}_\mathrm{MP}(s)$ of the model $\widetilde{\Sigma}(s)$, the IMC controller $Q(s)$ will be stable (as required) and implementing the full model $\widetilde{\Sigma}(s)$ as internal model yields the best model-plant match. However, since this approach does not invert the NMP part $\widetilde{\Sigma}_\mathrm{NMP}(s)$ of the model $\widetilde{\Sigma}(s)$, the closed-loop IMC behaviour $T(s)$ will still exhibit this NMP behaviour.

Design. Consider the q zeros of the model $\widetilde{\Sigma}(s)$ to be sorted such that the first ζ zeros $z_1, \ldots, z_\zeta$ are in the right half of the complex plane ($\mathrm{Re}(z_1), \ldots, \mathrm{Re}(z_\zeta) > 0$) and the other $q - \zeta$ zeros to be in the left half of the complex plane.

The following introduces two specific possibilities to split an NMP model $\widetilde{\Sigma}(s)$ into an MP $\widetilde{\Sigma}_\mathrm{MP}(s)$ and an NMP part $\widetilde{\Sigma}_\mathrm{NMP}(s)$:

$$\widetilde{\Sigma}(s) = \widetilde{\Sigma}_\mathrm{MP}(s) \cdot \widetilde{\Sigma}_\mathrm{NMP}(s) \quad \text{with either}$$

$$\text{Case I:} \quad \widetilde{\Sigma}_{\text{NMP}}(s) = \frac{\left(\frac{s}{z_1}+1\right)\cdot\ldots\cdot\left(\frac{s}{z_\zeta}+1\right)}{\left(\frac{s}{-z_1}+1\right)\cdot\ldots\cdot\left(\frac{s}{-z_\zeta}+1\right)} \triangleq \widetilde{\Sigma}_{\text{A}}(s), \quad \text{or} \tag{2.25}$$

$$\text{Case II:} \quad \widetilde{\Sigma}_{\text{NMP}}(s) = \frac{\left(\frac{s}{z_1}+1\right)\cdot\ldots\cdot\left(\frac{s}{z_\zeta}+1\right)}{\left(\frac{s}{\lambda}+1\right)^\zeta} \tag{2.26}$$

Both cases share the properties that the steady-state gain of the NMP part equals to one ($\widetilde{\Sigma}_{\text{NMP}}(0) = 1$) and that the relative degree of the MP part $\widetilde{\Sigma}_{\text{MP}}(s)$ is equal to the relative degree r of the original model $\widetilde{\Sigma}(s)$.

The design of an IMC controller $Q(s)$ holds for both cases and is analogous to the one portrayed for MP systems as introduced in Eq. (2.10)

$$\boxed{Q(s) = \widetilde{\Sigma}_{\text{MP}}^{-1}(s)F(s)} \tag{2.27}$$

with the IMC filter $F(s)$ from Eq. (2.10) and is shown as a block diagram in Fig. 2.10.

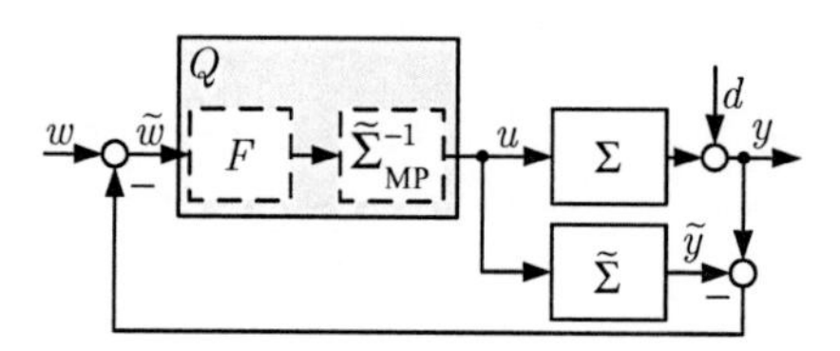

Fig. 2.10: IMC structure with IMC controller Q from Eq. (2.27).

Explanation. The resulting IMC controller $Q(s)$ is stable since it is obtained by inverting the MP part $\widetilde{\Sigma}_{\text{MP}}(s)$ of the model. Thus, nominal stability follows. Considering robust stability, one gets with the multiplicative output uncertainty $\Delta(s)$ from Eq. (2.9) with Property 2.4 on page 25 that if

$$\|\Sigma_0(s)\|_\infty = \|\Delta(s)\widetilde{\Sigma}_{\text{NMP}}(s)F(s)\|_\infty < 1 \tag{2.28}$$

holds then the closed IMC loop is stable. Thus, given an uncertain plant, the choice of the NMP part $\widetilde{\Sigma}_{\text{NMP}}(s)$ of the model will influence robust stability.

Since the IMC filter $F(s)$ has a steady-state gain of one ($|F(0)| = 1$), it follows from $Q(0) = \widetilde{\Sigma}_{\mathrm{MP}}(0)^{-1} = \widetilde{\Sigma}(0)^{-1}$ and Property 2.3 that the closed IMC loop yields zero steady-state offset.

Considering the closed-loop behaviour, one finds from Fig. 2.10 that

$$T(s) = \frac{y(s)}{w(s)} = \frac{F(s)\widetilde{\Sigma}_{\mathrm{NMP}}(s)\,(I - \Delta(s))}{I + F(s)\widetilde{\Sigma}_{\mathrm{NMP}}(s)\Delta(s)} \tag{2.29}$$

holds. Hence, even for small model uncertainties $\Delta(s)$ the closed-loop will retain the NMP behaviour $\widetilde{\Sigma}_{\mathrm{NMP}}(s)$ of the plant $\widetilde{\Sigma}(s)$:

$$\lim_{\Delta(s)\to 0} T(s) = F(s)\widetilde{\Sigma}_{\mathrm{NMP}}(s), \quad \forall s. \tag{2.30}$$

from which one finds

$$y(s) = F(s)\widetilde{\Sigma}_{\mathrm{NMP}}(s)w(s) + \left(1 - F(s)\widetilde{\Sigma}_{\mathrm{NMP}}(s)\right) d(s) \tag{2.31}$$

as the behaviour of the output signal $y(s)$. This, however, is to be expected since (with the demand of internal stability) an NMP behaviour of a plant cannot be influenced by feedback or feedforward by any control algorithm [63, 87].

Interpretation. *Case I:* The model $\widetilde{\Sigma}(s)$ is split into an MP part $\widetilde{\Sigma}_{\mathrm{MP}}(s)$ and a stable all-pass part $\widetilde{\Sigma}_{\mathrm{A}}(s)$. The all-pass part $\widetilde{\Sigma}_{\mathrm{A}}(s)$ has a gain of one over all frequencies $|\widetilde{\Sigma}_{\mathrm{NMP}}(j\omega)| = 1$, $\forall\omega$ and contains all right half plane zeros. Thus, the amplitude-over-frequency shape of the MP part $\widetilde{\Sigma}_{\mathrm{MP}}(s)$ is equal to that of the original model $\widetilde{\Sigma}(s)$:

$$|\widetilde{\Sigma}(j\omega)| = |\widetilde{\Sigma}_{\mathrm{MP}}(j\omega)|, \quad \forall\omega. \tag{2.32}$$

It is important to appreciate that a number of ζ zeros are *introduced* to the remaining MP part $\widetilde{\Sigma}_{\mathrm{MP}}(s)$ in Eq. (2.25) at the mirror image of the NMP zeros (namely at $-z_1, \ldots, -z_\zeta$). Thus, the system $\widetilde{\Sigma}_{\mathrm{MP}}(s)$ will have the same number of zeros and the same relative degree as the original non minimum-phase model $\widetilde{\Sigma}(s)$.

It can be shown [63, 72], that by splitting the model according to case I and choosing an infinitely fast filter pole $\lambda \to \infty$, the resulting IMC controller $Q(s)$ from Eq. (2.10) will minimise the integral square error (ISE) norm, i. e.,

$$\min_{Q(s)} \int_0^\infty |w(t) - y(t)|^2 dt. \tag{2.33}$$

Thus, $Q(s)$ from Eqns. (2.10) and (2.25) (case I) is ISE-optimal for $\lambda \to \infty$.

Case II: Case II differs from case I in the choice of the ζ zeros that are introduced to the MP part $\widetilde{\Sigma}_{\mathrm{MP}}(s)$. The property of Eq. (2.33) is interesting but not necessary for IMC design. Thus, since ζ zeros need to be introduced to the MP part $\widetilde{\Sigma}_{\mathrm{MP}}(s)$ it is convenient to place them at the same location as the IMC filter pole, namely at $-\lambda$.

The ISE-optimal IMC controller $Q(s)$ is interesting from a mathematical perspective but not necessarily a good choice for practical applications.

Aside from the unrealistic necessity of using an infinitely fast IMC filter pole (one could use an arbitrary but finite "fast" pole), such a design might lead to excessive noise amplification. Foremost, however, an integral criterion does not account for the shape of a control response since it merely represents its norm. Thus, ISE-optimal controllers may lead to oscillating control inputs $u(t)$ and oscillating plant outputs $y(t)$ which result in high wear and tear of the actuator. Moreover, a control response is judged intuitively by an engineer, who tries to avoid oscillating input and output signals as they are not considered good responses. Therefore, the author encourages the designer to choose case II since it offers more freedom in designing the controller.

The following example demonstrates the difference between an IMC design for non-minimum phase systems using either Case I or Case II.

Example 2.2 (IMC control of a linear non-minimum phase process):
This example has been adapted from [9]. Consider the non-minimum phase exact model

$$\widetilde{\Sigma}(s) = \Sigma(s) = \frac{0.225s^2 - 0.0964s + 1}{0.137s^3 + 1.274s^2 + 2.137s + 1} \tag{2.34}$$

with complex conjugate zeros at $z_{1/2} = 0.21 \pm j2.1$, a two-fold pole at $p_1 = p_2 = -1$, and a single pole at $p_3 = -7.3$. Hence, the plant is stable and NMP. With case I, one finds for an ISE-optimal controller $Q_{\mathrm{ise}}(s)$ with Eqns. (2.14b),(2.25) and (2.27)

$$Q_{\mathrm{ise}}(s) = \frac{0.137s^3 + 1.274s^2 + 2.137s + 1}{0.225s^2 + 0.0964s + 1} \cdot \frac{1}{s/\lambda + 1}. \tag{2.35}$$

With case II (cf. Eq. (2.26)), one finds with Eqns. (2.14b) and (2.27):

$$Q_{\lambda}(s) = \frac{0.137s^3 + 1.274s^2 + 2.137s + 1}{(s/\lambda + 1)^2} \cdot \frac{1}{s/\lambda + 1}. \tag{2.36}$$

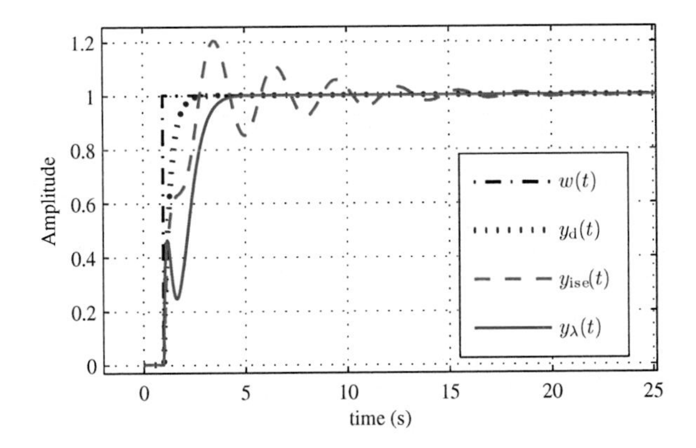

Fig. 2.11: Step response of Example 2.2.

Both controllers were designed using $\lambda = 3$. As the filter parameter λ is not chosen to be infinite, Q_{ise} will be referred to as ISE-oriented, henceforth in the example. Figure 2.11 shows the step response of the plant $\Sigma(s)$ controlled with both the ISE-oriented IMC controller $Q_{\text{ise}}(s)$ (output $y_{\text{ise}}(t)$) and IMC controller $Q_\lambda(s)$ (output $y_\lambda(t)$). The ISE-oriented controller $Q_{\text{ise}}(s)$ shows an oscillating output $y_{\text{ise}}(t)$, since the NMP zeros were underdamped which leads to underdamped poles in the controller. In any automotive control system, such an output trajectory would be unacceptable due to the oscillations and the slow settling time of about 20 seconds. This step response yields an ISE (cf. Eq. (2.33)) of $\int_0^\infty |w(t) - y_{\text{ise}}(t)|^2 dt \approx 0.3595$.

The output $y_\lambda(t)$ of the closed-loop with the IMC controller $Q_\lambda(s)$, where the additional MP zeros were placed at the IMC filter pole $-\lambda$, shows considerably more damping and a settling time of about four seconds. To an engineer, the output $y_\lambda(t)$ is significantly more appealing than the ISE-oriented, although its ISE norm $\int_0^\infty |w(t) - y_\lambda(t)|^2 dt \approx 0.6214$ is almost twice as large.
■

In conclusion, IMC for NMP systems can be designed to be ISE-optimal, however, a low value of the ISE-norm usually does not stand in any relationship with the desired closed-loop behaviour. Therefore, an ISE-optimal control design is not of any interest in an automotive application.

2.2.3 *Input Constraints*

The most basic form of incorporating input constraints into IMC is briefly reviewed since this topic is of high importance in virtually all practical control applications.

Input constraints are given as

$$u_{\min} \leq u(t) \leq u_{\max}. \tag{2.37}$$

Since this is a property of the *plant*, one possibility of treating this limitation is to include it also in the model $\widetilde{\Sigma}$. Figure 2.12 shows the resulting

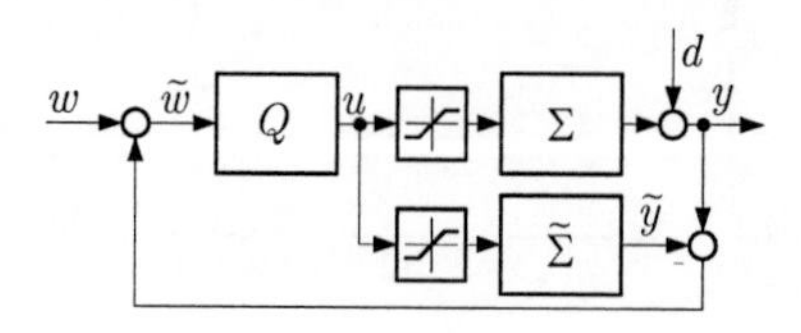

Fig. 2.12: Handling input constraints by including the limitation in the model $\widetilde{\Sigma}$.

IMC loop where the model is enhanced to respect input constraints.

This approach does not change robust stability since the modelling error is not affected. However, since the IMC controller $Q(s)$ does not have any information on limited inputs, performance will decrease. Other approaches to limit inputs with IMC are, for example, given in [9, 84, 91, 104]. These approaches yield better performance than the one portrayed here, but are mathematically more involved, which spoil the attractive simplicity of IMC design. To this end, Section 5.1 develops an effective and straightforward method of respecting input constraints with IMC.

2.3 Feasibility of IMC as Automotive Controller

IMC is a feasible control concept for automotive applications. In order for a control concept to be of interest to the automotive industry, it should comply with the demands D1-D5 (listed in Section 1.3.2) and it should be a model-based design procedure (described in Section 1.3.4 on pages 13ff). Since IMC is a model-based controller, it naturally fits in the model-based design procedure. The following shows that an IMC controller meets the requirements on automotive controllers.

Comparing the automotive requirements D1-D5 to the IMC design method.

D1: Performance criteria are given in time-domain. Time domain criteria are incorporated into an IMC design in a straightforward fashion by choosing an IMC filter $F(s)$ which meets these criteria. Considering minimum phase plants, the closed-loop behaviour $T(s)$ is identical to the IMC filter $F(s)$ (see Eq. (2.18)). Considering NMP plants, the closed-loop behaviour is governed by the IMC filter $F(s)$ together with the (unavoidable) NMP behaviour $\widetilde{\Sigma}_{\text{NMP}}(s)$ of the plant (see Eq. (2.29)). Hence, performance criteria in time-domain are accounted for by choosing the IMC filter $F(s)$, according to the desired closed-loop behaviour (and, if applicable, appreciating the unavoidable NMP behaviour of plants).

D2: Robust stability and zero steady-state offset are required. These conditions are basic properties of the IMC structure and are fulfiled automatically (see Section 2.1.3) if the design procedure, as introduced above, is followed.

D3: An IMC must be implemented in a car's OCU. An IMC does not rely on numerically intense calculations such as on-line matrix inversions or on-line optimisations. It can be assumed, that an IMC can be implemented in a standard OCU with a computational burden of the same order of magnitude as today's PID oriented automotive controllers.

From the design as introduced in the preceding sections, it follows that, since the model is part of the final controller $K(s)$ (see Fig. 2.1), a controller $K(s)$ implemented in the IMC structure will have an order of $2n$, where n is the order of the model. However, if an IMC is implemented in a standard control loop (see Fig. 2.4 and Eq. (2.3)), it follows that the minimal realisation of the controller $C(s)$ has an order of n. Hence, it is proposed to *always* implement an IMC controller as a standard controller $C(s)$.

D4: A controller must be straightforward to calibrate. Calibration of an IMC is significantly easier than calibration of a PID controller. An IMC can be calibrated by adjusting the parameters $b_0, \ldots, b_q$ and $a_0, \ldots, a_n$ (cf. Eq. (2.1)) of the internal model. The advantage of adjusting model parameters over adjusting controller parameters, like PID gains, is appealing to an engineer, especially if the model parameters have a physical interpretation. Moreover, model parameters can be calibrated on-line at a test bed without a time-intensive new design.

If desired, the filter parameter λ can also be calibrated. However, since this parameter defines the nominal closed-loop bandwidth and noise amplification, it can be determined from the closed-loop specifications, alone.

D5: A controller should be easy to understand. The IMC design as presented here relies on the basic properties of the IMC structure (see Section 2.1.3) which can be comprehended quickly.

In conclusion, the IMC concept fulfils all requirements of the automotive industry.

2.4 Summary

The design of internal model control focuses on finding a controller such that a given closed-loop I/O behaviour is achieved. It is based on feedforward control design and relies on model inversion. The IMC structure is simple and plausible and provides valuable properties such as nominal and robust stability as well as zero steady-state offset.

The attractiveness of IMC to industry comes from the internal model and the simple design law. Through this model the IMC controller $Q(s)$ is defined to a great extent. Once an IMC controller $Q(s)$ is determined for a specific plant $\Sigma(s)$, it can be adapted to control similar plants by calibrating the internal model parameters and – if desired – the tuning parameter λ. By changing the model parameters, the IMC controller parameters are changed accordingly. This enables non-control engineers to calibrate an existing IMC controller, since knowledge of the plant suffices to determine the model parameters. IMC does not use on-line optimisation procedures. Hence, it can be implemented in a real-time environment like an OCU.

In order to use the IMC structure to control nonlinear systems a nonlinear IMC controller Q for stable nonlinear systems is developed in the following.

3. INTERNAL MODEL CONTROL OF NONLINEAR SISO SYSTEMS

This chapter develops the theoretical contribution of this thesis, namely the extension of the SISO IMC principle to nonlinear SISO systems. The IMC design philosophy, including the IMC structure of the linear case as introduced in Chapter 2, is retained. Hence, a nonlinear IMC is designed as a feedforward controller for the nonlinear system by employing a model inverse in connection with a low-pass filter.

This chapter uses basic system theoretical properties, like the relative degree, and basic functional analytic ideas as a right inverse, operators and signal norms to propose a control design method. The appeal of the proposed design method is that it is applicable to industrial control problems, its application is straightforward, and, despite its simplicity, offers important properties such as nominal and robust stability.

3.1 *Mathematical Preliminaries*

For simplicity, the concept of nonlinear IMC is presented for the SISO case, only. This chapter uses a functional analytic view on dynamical systems, where the input/output (I/O) behaviour of the system $\widetilde{\Sigma}$ is given by understanding $\widetilde{\Sigma}$ as an operator which maps a signal u contained in a function space $\mathcal{U}$ into a signal $\tilde{y}$, contained in the function space $\widetilde{\mathcal{Y}}$, i. e.,

$$\widetilde{\Sigma} : \mathcal{U} \to \widetilde{\mathcal{Y}}. \tag{3.1}$$

It is written as $\tilde{y} = \widetilde{\Sigma} u$, for a $u \in \mathcal{U}$. The usual composition symbol "$\circ$" is omitted in this text, except for instances where the composition is emphasised. In its state space representation, the model $\widetilde{\Sigma}$ is given as

$$\widetilde{\Sigma} : \; \dot{\boldsymbol{x}}(t) = \boldsymbol{f}\left(\boldsymbol{x}(t), u(t)\right), \quad \boldsymbol{x}(0) = \boldsymbol{x}_0, \; \boldsymbol{x} \in \mathcal{X}, \tag{3.2a}$$

$$\tilde{y}(t) = h\left(\boldsymbol{x}(t), u(t)\right), \quad u \in \mathcal{U}, \; \tilde{y} \in \widetilde{\mathcal{Y}}. \tag{3.2b}$$

The signals $u(t) \in \mathbb{R}^1$, $\boldsymbol{x}(t) \in \mathbb{R}^n$ and $y(t) \in \mathbb{R}^1$ denote values at a specific time point t. The trajectory of a signal is denoted by omitting the dependence of t (e. g., $\boldsymbol{x}, \tilde{y}$) and refers to the input $u \in \mathcal{U}$, state $\boldsymbol{x} \in \mathcal{X}$ or output $\tilde{y} \in \widetilde{\mathcal{Y}}$ as the whole time function. Thus, instead of using the expression of, for example, $\boldsymbol{x}(\cdot)$ to denote a trajectory, this work omits "$(\cdot)$". The time t is defined on the set $t \in \mathcal{T} = [0, \infty)$. Moreover, the model $\widetilde{\Sigma}$ is assumed to be time-invariant and asymptotically[1] stable.

The vector field $\boldsymbol{f}$ and the function h are analytic[2] in their arguments $\boldsymbol{x}(t)$ and $u(t)$ for all $\boldsymbol{x} \in \mathcal{X}$ and $u \in \mathcal{U}$. Thus, the solution $\boldsymbol{x}$ of Eq. (3.2a) exists and is unique [46]. Moreover, analycity implies that the functions $\boldsymbol{f}$ and h are non-singular with respect to all possible arguments $\boldsymbol{x}(t)$ and $u(t)$. The measurement map $h : \mathbb{R}^n \times \mathbb{R}^1 \to \mathbb{R}^1$ maps the current values $\boldsymbol{x}(t)$ and $u(t)$ of the states and input signal into the current value $y(t)$ of the output signal. Note that the function h in Eq. (3.2b) may or may not be directly dependent on u. If the input u appears in the mapping h explicitly, the model is said to have a direct feedthrough.

The initial state $\boldsymbol{x}_0$ of the system is assumed to be given and fixed, which simplifies the following notation in such that a system can be represented by a single operator $\widetilde{\Sigma}$. If the initial condition $\boldsymbol{x}_0$ would be allowed to vary, one would have to introduce a different operator for each initial condition or use the concept of relations instead of operators [99].

The output-function space

$$\widetilde{\mathcal{Y}} = \left\{ \tilde{y} \in \widetilde{\mathcal{Y}} : \ \tilde{y} = \widetilde{\Sigma} u, \ u \in \mathcal{U} \right\} \tag{3.3}$$

contains exclusively all signals which can be produced by the model $\widetilde{\Sigma}$ under the given initial state $\boldsymbol{x}_0$ and permissible controls u. Hence, the I/O map $\widetilde{\Sigma}$ is *surjective* on this set.

In the following, the identity operator I will be used extensively and is defined as

$$u = Iu \quad \text{for any trajectory } u. \tag{3.4}$$

Thus, the identity operator preserves its input signal.

In order to define the system gain, signal norms are introduced. The

[1] Asymptotic stability (see e. g., [54, 83]) is a property of the behaviour of the states of a system. Since this thesis is mainly concerned with I/O behaviour, stability conditions in dependence of the I/O behaviour are given in the following.

[2] An analytic function (see e. g., [16, 56] for its properties) is also referred to as "holomorphic function."

norm $\mathcal{L}_p$ of a signal u is defined as

$$\|u\|_{\mathcal{L}_p} = \left(\int_{t \in \mathcal{T}} |u|^p dt \right)^{\frac{1}{p}} \tag{3.5}$$

with $1 \leq p < \infty$. A signal u for which the norm $\|u\|_{\mathcal{L}_p}$ exists, is denoted by $u \in \mathcal{L}_p$. For $p = \infty$ the $\mathcal{L}_\infty$-norm is

$$\|u\|_{\mathcal{L}_\infty} = \sup_{t \in \mathcal{T}} |u(t)| < \infty. \tag{3.6}$$

In the following, a signal norm is denoted by $\|\cdot\|$ whenever the dependence on p is not fixed.

Remark 3.1. The following expressions are used synonymously:

- $u \in \mathcal{L}_\infty$
- u is bounded piecewise continuous (cf. [54, 99])
- u is non-explosive (cf. [99])
- u is measurable

The above all mean that the signal $u(t)$ is defined at each instance in time, i. e., $\exists\, u(t),\ \forall t \in \mathcal{T}$.

Definition 3.1 (Finite-Gain Stability [54]**).** An I/O map $\widetilde{\Sigma}$ is called to be finite-gain $\mathcal{L}_p$-stable if there exist non-negative constants γ and β such that

$$\|\widetilde{\Sigma} u\| \leq \gamma \|u\| + \beta \tag{3.7}$$

holds for all $u \in \mathcal{L}_p$.

◇

In this work, stability means finite-gain stability unless noted differently.

Definition 3.2 (System gain, [54]**).** For a finite-gain $\mathcal{L}_p$-stable system, the smallest value γ, for which inequality (3.7) is satisfied, is called the gain of the system, denoted by $g(\widetilde{\Sigma}) = \gamma$. ◇

Remark 3.2. If the operator $\widetilde{\Sigma}$ has the property

$$0 = \widetilde{\Sigma}\, 0 \tag{3.8}$$

where 0 is a trajectory $u(t) = 0,\ \forall t$ then the system gain can also be expressed as [99]

$$g(\widetilde{\Sigma}) = \sup \frac{\|(\widetilde{\Sigma} u)\|}{\|u\|} \tag{3.9}$$

where the supremum is taken over all $u \in \mathcal{U}$ for which $u \neq 0$.

If the operator $\widetilde{\Sigma}$ does not have the property $0 = \widetilde{\Sigma}\, 0$ it can be enforced by adding a compensating bias β (cf. Eq. (3.7)) to the output $\tilde{y}$.

Gains share the properties of norms. In addition [99], a gain fulfils the inequality

$$g(A \circ B) \leq g(A) g(B) \tag{3.10}$$

for any two stable operators A and B.

Lie-derivative. The following relationships make use of the Lie derivative [51] to simplify notation. The time derivative of $h(\boldsymbol{x})$ along $\boldsymbol{f}(\boldsymbol{x}, u)$ is denoted by

$$L_f h(\boldsymbol{x}) = \frac{\partial h(\boldsymbol{x})}{\partial \boldsymbol{x}} \boldsymbol{f}(\boldsymbol{x}, u)$$

and is equal to $\dot{\tilde{y}} = \frac{d}{dt} h(\boldsymbol{x})$. The function $L_f^k h(\boldsymbol{x})$ satisfies the recursion:

$$L_f^k h(\boldsymbol{x}) = \frac{\partial L_f^{k-1} h(\boldsymbol{x})}{\partial \boldsymbol{x}} \boldsymbol{f}(\boldsymbol{x}, u),$$

with $L_f^0 h(\boldsymbol{x}) = h(\boldsymbol{x})$.

Remark 3.3. Consider the Lie-derivative of a function h as a time derivative of that function given in dependence upon the system states $\boldsymbol{x}$. It is an abbreviation for the chain-rule with the subsequent substitution of $\dot{\boldsymbol{x}}$ by its definition Eq. (3.2a).

In this work, the expressions $L_f^k h(\boldsymbol{x}, u)$ and $L_f^k h(\boldsymbol{x})$ are used synonymously, despite the omitted dependency on u. In the case of the explicit dependency on u, the Lie-derivative has to be defined by

$$L_f h(\boldsymbol{x}, u) = \frac{\partial h(\boldsymbol{x}, u)}{\partial \boldsymbol{x}} \boldsymbol{f}(\boldsymbol{x}, u) + \frac{\partial h}{\partial u} \dot{u} \quad \text{with the recursion}$$

$$L_f^k h(\boldsymbol{x}, u) = \frac{\partial L_f^{k-1} h(\boldsymbol{x}, u)}{\partial \boldsymbol{x}} \boldsymbol{f}(\boldsymbol{x}, u) + \sum_{i=1}^{k-1} \frac{\partial L_f^i h(\boldsymbol{x}, u)}{\partial u^{(i-1)}} u^{(i)}$$

$$\text{and } L_f^0 h(\boldsymbol{x}, u) = h(\boldsymbol{x}, u).$$

Fortunately, derivations of the output function h which contain derivatives of u will never be used in this work. In this light, the definition given in this remark is never employed and merely given for completeness.

Definition 3.3 (Relative degree [35, 42]). The relative degree r of a system $\widetilde{\Sigma}$ is the smallest value $r \in \mathbb{N}_0$ for which $\tilde{y}^{(r)}$ can be expressed by an algebraic function φ which *explicitly* depends on the input u, i. e.,

$$\tilde{y}^{(r)} = L_f^r h(\boldsymbol{x}, u) = \varphi(\boldsymbol{x}, u) \tag{3.11}$$

holds with

$$\frac{\partial}{\partial u} L_f^i h(\boldsymbol{x}, u) = 0, \quad 0 \leq i \leq r - 1$$

$$\frac{\partial}{\partial u} L_f^r h(\boldsymbol{x}, u) \neq 0.$$

If the relative degree r does not exist (i. e., an equality (3.11) cannot be given), it will be referred to as $r \to \infty$ (see e. g., [19, 93], where this convention is also used).

The relative degree is *well-defined* if the value of r is constant in the state-space region of concern (i. e., locally or globally, depending on the context). It is called *ill-defined* if it is not well-defined. ◇

Remark 3.4. The relative degree r can also be defined more condensed by

$$r = \arg\min_k \left\{ \frac{\partial}{\partial u} L_f^k h(\boldsymbol{x}, u) \neq 0 \right\}.$$

It is sometimes also called *relative order* and can be interpreted as the number of integrations that the input u (or some algebraic function of it) has to undergo until it affects the output $\tilde{y}$. Therefore, r is an indication of the *sluggishness* of the system response [19]. For the relative degree of $r = 0$, the system has a direct feedthrough.

This definition coincides with Definition 2.1 of the relative degree r of a transfer function: By multiplying the transfer function with s^r (which is equivalent to taking r derivatives) one obtains a transfer function with direct feedthrough.

An excellent, detailed, structural interpretation of the relative degree r and its implications can be found in [19]. The work of [19] is strongly recommended to the interested reader, as its results complement the definition and interpretation of the relative degree given here. For the results of this thesis, the notion of relative degree is of fundamental importance and the relationship (3.11) will be used extensively.

3.2 Output-Function Space

This thesis mainly focuses on the input/output behaviour of nonlinear systems by using the abstract notion of operators, as in Eq. (3.1). Within this context, the definition of the input space $\mathcal{U}$ and the output space $\widetilde{\mathcal{Y}}$ of the operator (plant model) $\widetilde{\Sigma}$ becomes necessary. In the nonlinear IMC design, as it is proposed in the following, the precise knowledge of the input-space $\mathcal{U}$ and, more importantly, the output-space $\widetilde{\mathcal{Y}}$ is necessary for more than just the sake of completeness of the mathematical concept of operators. These spaces, describing the *shapes* of the input function u and the resulting output-function $\tilde{y}$, are used for control design and their knowledge is necessary for e. g., respecting input constraints.

Remark 3.5. Note that, here, the word *shape* is used to distinguish the goal of describing the output-function space $\widetilde{\mathcal{Y}}$ from the usually encountered definition of the output-function space $\widetilde{\mathcal{Y}}$ by means of a signal norm. In loose terms, it means that the look of all output functions $\tilde{y} \in \widetilde{\mathcal{Y}}$ is to be described by their initial behaviour at time $t = 0$ as well as how many times they are (at least) continuously differentiable in the development for $t > 0$. The number of times a function is continuously differentiable can be interpreted as its smoothness.

Throughout this thesis, the input space $\mathcal{U}$ is to be regarded as the space of all piecewise continuous functions with constant constraints $u_{\min}$, $u_{\max}$, i. e.,

$$\mathcal{U}: \quad \{u | \; u \in \mathcal{L}_\infty : u_{\min} \leq u(t) \leq u_{\max}\}, \quad u_{\min}, u_{\max} \in \mathbb{R}. \qquad (3.12)$$

In general, it is not necessary for the constraints $u_{\min}$, $u_{\max}$ to be constant. However, this demand sufficiently describes the typical scenario of control problems and simplifies later results.

The set $\widetilde{\mathcal{Y}}$, which contains all *possible* output functions $\tilde{y}$ of an operator $\widetilde{\Sigma}$, is dependent on the structure of $\widetilde{\Sigma}$ (in particular upon its relative degree r) and the possible input functions $u \in \mathcal{U}$. In the following, the output-function space $\widetilde{\mathcal{Y}}$ is defined in two propositions. Proposition 3.1 addresses the initial (i. e., at $t = 0$) behaviour of all functions $\tilde{y} \in \widetilde{\mathcal{Y}}$ and Proposition 3.2 defines the output function space $\widetilde{\mathcal{Y}}$ from $t > 0$ in dependence upon the models relative degree r and the input function space $\mathcal{U}$.

The results obtained below present a generalisation of the results obtained by [19]. It will be shown that, for a nonlinear system (3.2) with relative degree r, the output trajectory $\tilde{y}$ has a fixed initial shape and its smoothness can be defined by the system's relative degree r.

Proposition 3.1 (Initial shape of all output functions in $\widetilde{\mathcal{Y}}$). *Let $\widetilde{\Sigma}$ be a model as given by Eq. (3.2) with the initial state $\boldsymbol{x}_0$ and the relative degree r in some (arbitrarily small) neighbourhood around $\boldsymbol{x}_0$. The set $\widetilde{\mathcal{Y}}$ contains* only *such signals $\tilde{y}$ that have an initial shape (i. e., at time $t = 0$) given by*

$$\begin{aligned}\tilde{y}(0) &= h(\boldsymbol{x}_0)\\ \dot{\tilde{y}}(0) &= \dot{h}(\boldsymbol{x}_0) = L_f h(\boldsymbol{x}_0)\\ &\vdots\\ \tilde{y}^{(r-1)}(0) &= L_f^{r-1} h(\boldsymbol{x}_0).\end{aligned} \tag{3.13}$$

Proof. Equations (3.13) follow directly from differentiations of the output map $h(\boldsymbol{x})$ (cf. Eq. (3.2b)) under initial conditions $\boldsymbol{x}_0$, given in Eq. (3.2a). □

Note that the initial shape of the output function $\tilde{y}$ up to its $r-1$-th derivative $\tilde{y}^{(r-1)}$ is not influenced by the initial control $u(0)$. Rather, it is defined completely by the initial state $\boldsymbol{x}_0$. In [69], this initial behaviour is investigated from a differential-algebraic perspective and provides additional inside into the matter.

Proposition 3.2 (Output-function space of a system with a well-defined relative degree r). *Consider a model $\widetilde{\Sigma}$ given in Eq. (3.2) with a well-defined relative degree $r \in \mathbb{N}$.*
Then, the output function space $\widetilde{\mathcal{Y}}$ can be defined as follows[a]*: At a given state signal $\boldsymbol{x}$, the set $\widetilde{\mathcal{Y}}$ contains* only *signals $\tilde{y}$ that have the initial shape as defined in Proposition 3.1 where the possible following shape (i. e., $t \geq 0$) is defined by*

$$\widetilde{\mathcal{Y}} = \left\{ \tilde{y} \mid \tilde{y} \subseteq C^{r-1} : \tilde{y}^{(r)} \in \{\varphi(\boldsymbol{x}, u) : u \in \mathcal{U}\} \right\}. \tag{3.14}$$

[a] Note that the expression C^k means the space of all k-times continuously differentiable functions.

In words, the shape of all output functions $\tilde{y} \in \widetilde{\mathcal{Y}}$ is defined via the range of the r-th derivative $\tilde{y}^{(r)} = \varphi(\boldsymbol{x}, u)$ (cf. Definition 3.3) that is reachable from the input $u \in \mathcal{U}$.

The proof extensively uses some of the basic properties of (real-)analytic functions. Those are the following (see e. g., [16, 56]):

- Analytic functions are infinitely often continuously differentiable (i. e., they are C^∞).
- The Taylor series of an analytic function converges.
- Products, sums and compositions of analytic functions are also analytic.
- A division $\frac{f}{g}$ of two analytic functions f and g is also an analytic function except for the point where the singularity occurs (i. e., at $g = 0$).

Remark 3.6. The proof of the Proposition above could be briefly given as "The proof follows directly from the analycity of $\boldsymbol{f}$ and h". Such a wording is usually encountered in mathematical treatments on similar issues. Here, however, the proof is given in detail as the author believes that some explanations are helpful in understanding this matter.

Proof. The proof is given in three steps. First, it is shown that if r exists then $\varphi(\boldsymbol{x}, u)$ exists (in the sense that it has a finite value at all times $t \geq 0$):

By definition, both the output map $h(\boldsymbol{x}, u)$ and the vector field $\boldsymbol{f}(\boldsymbol{x}, u)$ are analytic in $\boldsymbol{x}$ and u. Therefore, the function $\varphi(\boldsymbol{x}, u)$, introduced in Eq. (3.11), is also an analytic function of its arguments $\boldsymbol{x}$ and u since it is obtained only by differentiations, products and sums of the analytic functions h and $\boldsymbol{f}$ (see Definition 3.3). As $\varphi(\boldsymbol{x}, u)$ is analytic, it is non-singular for all permissible $\boldsymbol{x}$ and u and, thus, has a finite value at all times $t \geq 0$. Therefore, $\varphi(\boldsymbol{x}, u)$ exists if r exists.

As an intermediate step, it is shown that $\varphi(\boldsymbol{x}, u)$ is integrable in time: The function $\varphi(\boldsymbol{x}, u)$ is dependent on $\boldsymbol{x}$ and u. The state trajectory $\boldsymbol{x}$ is an n-vector of continuous functions of time t ($x_i \in C^0$) and the input u is, in the most general case, discontinuous in time t ($u \in \mathcal{L}_\infty$). Both exist at all times. It follows from the chain-rule that the composition of an analytic function (here φ) with a piecewise continuous $\mathcal{L}_\infty$- (or C^k-) function is also a $\mathcal{L}_\infty$- (or C^k-) function (with $k \in \mathbb{N}$). Therefore, $\varphi(\boldsymbol{x}, u)$ can generally be described as a piecewise continuous function *of time* t since u can – in the worst case – be discontinuous *in time*:

$$\varphi(\boldsymbol{x}, u) = \tilde{y}^{(r)} \in \mathcal{L}_\infty. \tag{3.15}$$

As $\varphi(\boldsymbol{x}, u)$ is piecewise continuous, it is integrable.

Finally, the proof of Eq. (3.14) is completed by showing that the output signal $\tilde{y}$ is $r-1$ times continuously differentiable (i. e., $\tilde{y} \in C^{r-1}$): From Definition 3.3 and Eq. (3.15) one finds

$$\tilde{y}(t) = \underbrace{\int_T \cdots \int_T}_{r} \varphi(\boldsymbol{x}(\tau), u(\tau)) d\tau^r \ + c, \quad \text{where } c \in \mathbb{R}. \tag{3.16}$$

c is some constant that can be obtained from Proposition 3.1. It follows that $\tilde{y}$ is r-times differentiable, since it can be obtained by r integrations. □

Interpretation. The implications of Proposition 3.2 are briefly discussed. Consider linear systems and therein especially low-pass filters of orders from zero to three, e. g.,

$$F_0(s) = \frac{1}{1}, \ F_1(s) = \frac{1}{s+1}, \ F_2(s) = \frac{1}{(s+1)^2}.$$

Their respective relative degree is $0, 1, 2$. Using Proposition 3.2, it is possible to explain the well-known behaviour of these systems. Systems,

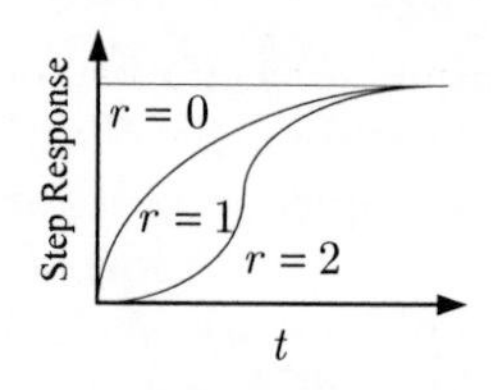

Fig. 3.1: Output shape of systems $F_0(s)$, $F_1(s)$ and $F_2(s)$ with relative degree r under the input discontinuity of a step input

as $F_0(s)$, which have a direct feedthrough, will react with a discontinuous output trajectory $\tilde{y}$ to a discontinuous input u. The first order filter $F_1(s)$, having an integrator between input and output, can only produce continuous output trajectories $\tilde{y}$ but will produce a discontinuous first derivative $\dot{\tilde{y}}$ upon a control discontinuity. Similarly, $F_2(s)$ will always produce at least continuous first derivatives $\dot{\tilde{y}}$ but discontinuous second derivatives $\ddot{\tilde{y}}$ following a discontinuous input u. This behaviour is shown in Fig. 3.1. Note, that the above considerations do not hold when u is chosen as a Dirac impulse, as this input will "bypass" one integrator. However, such an input is prohibited, as it does not lie in the set of $\mathcal{L}_\infty$.

As a main result, if follows from Proposition 3.2 that this observation about system behaviour also holds for nonlinear systems with a well-defined relative degree r. Thus, the relative degree r describes a structural property of dynamic systems, namely the *minimal smoothness of the system output.* The word "minimal" refers to the case when the input trajectory u is chosen to be discontinuous (i. e., $u \in \mathcal{L}_\infty$ and $u \notin C^0$). The following corollary addresses the question as to how the shape $\widetilde{\mathcal{Y}}$ of the output $\tilde{y}$ changes if u is chosen to have a certain smoothness itself ($\mathcal{U} \subseteq C^k$ with $k \in \mathbb{N}$).

Corollary 3.1 (System output space $\widetilde{\mathcal{Y}}$ under smooth inputs $u \in C^k$ and well-defined relative degree). *Consider a model $\widetilde{\Sigma}$ given in Eq.* (3.2) *with a well-defined relative degree $r \in \mathbb{N}$. Further, the input-function space is $\mathcal{U} \subseteq C^k$ with $k \in \mathbb{N}$.*
Then, the output function space $\widetilde{\mathcal{Y}}$ can be defined as follows: At a given state signal $\boldsymbol{x}$, the set $\widetilde{\mathcal{Y}}$ contains only *signals $\tilde{y}$ that have the initial shape as defined in Proposition 3.1 where the possible following shape (i. e., $t \geq 0$) is defined by*

$$\widetilde{\mathcal{Y}} = \left\{ \tilde{y} \mid \tilde{y} \subseteq C^{r-1+k} : \tilde{y}^{(r)} \in \{\varphi(\boldsymbol{x}, u) : u \in \mathcal{U}\} \right\}. \tag{3.17}$$

In loose terms, the implication of the above corollary is that the smoothness of the input u is "added" to the minimal smoothness of the output $\tilde{y}$.

Finally, the following addresses the case where the relative degree r is ill-defined (i. e., its value may change abruptly in time). The result is obtained from a piecewise application of Proposition 3.2 for time periods where the relative degree is well-defined. Consider a model $\widetilde{\Sigma}$ given in Eq. (3.2) with an ill-defined relative degree $r(t) \in \mathbb{N}$ or $r(t) \to \infty$. The series of values $r(t)$ over time can, thus, be constant or non-existent over a time interval. Then, the output function $\tilde{y}$ belongs piecewise to sets $\widetilde{\mathcal{Y}}_t$ as follows: If $r(t)$ is finite and constant within $t_1 \leq t < t_2$ then $\tilde{y}(t) \in \widetilde{\mathcal{Y}}_t$ can be defined within t by applying Proposition 3.2 to obtain $\widetilde{\mathcal{Y}}_t$.

If the relative degree is non-existent (i. e., $r \to \infty$) then $\tilde{y}(t) \in \widetilde{\mathcal{Y}}_t$ is an analytic function within t. $\tilde{y}(t)$ can then be given by its Taylor series

$$\begin{aligned} \tilde{y}(\tau) =& h(\boldsymbol{x}(t_1)) + L_f h(\boldsymbol{x}(t_1)) \cdot (\tau - t_1) + \frac{L_f^2 h(\boldsymbol{x}(t_1))}{2!} \cdot (\tau - t_1)^2 \\ & + \ldots + \frac{L_f^k h(\boldsymbol{x}(t_1))}{k!} \cdot (\tau - t_1)^k \quad \text{for} \quad k \to \infty. \end{aligned} \tag{3.18}$$

Let

$$r_{\min} = \min_t r(t), \quad \forall t \in [0, \infty). \tag{3.19}$$

The set $\widetilde{\mathcal{Y}}$ can be defined as containing signals $\tilde{y}$ with an initial behaviour $\tilde{y}(0)$ as defined by Proposition 3.1 with the following behaviour (i. e., $t \geq 0$) given by

$$\widetilde{\mathcal{Y}} \subseteq C^{r_{\min}-1}. \tag{3.20}$$

The main results of the above are that if the relative degree does not exist (i. e., $r \to \infty$) then the output trajectory $\tilde{y}$ is infinitely smooth and not influenced by the system input u and that the minimal smoothness of the system output $\tilde{y}$ depends on the smallest value $r_{\min}$ that the relative degree r possesses in time.

3.3 Right Inverse of the Model

This section introduces the properties of a right inverse of a dynamical SISO system. The notion of the right inverse is of high importance in the design of a nonlinear IMC controller. However, at this point it will not be discussed how a right inverse of a given dynamical system can be obtained but rather how it is defined. In Sections 4.1 and 4.2 right inverses for different system classes are given.

Definition 3.4 (Right inverse [28]). The right inverse

$$\widetilde{\Sigma}^{\mathrm{r}} : \widetilde{\mathcal{Y}} \rightarrow \mathcal{U} \tag{3.21}$$

of the system $\widetilde{\Sigma}$ is a mapping with the property

$$\widetilde{\Sigma}\widetilde{\Sigma}^{\mathrm{r}}\tilde{y} = \tilde{y} \tag{3.22}$$

for all $\tilde{y} \in \widetilde{\mathcal{Y}}$. ◇

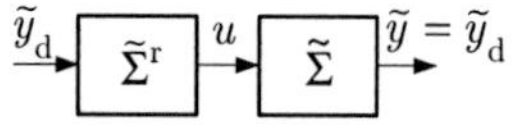

Fig. 3.2: Right inverse.

Thus, for every given signal $\tilde{y}_{\mathrm{d}} \in \widetilde{\mathcal{Y}}$, the right inverse $\widetilde{\Sigma}^{\mathrm{r}}$ generates an input u such that the model output $\tilde{y}$ exactly follows the trajectory $\tilde{y}_{\mathrm{d}}$ (Fig. 3.2). The domain $\widetilde{\mathcal{Y}}$ of the right inverse $\widetilde{\Sigma}^{\mathrm{r}}$ is equal to the range $\widetilde{\mathcal{Y}}$ of the plant model $\widetilde{\Sigma}$. According to Eq. (3.21), the right inverse is only applicable to signals $\tilde{y}_{\mathrm{d}}(= \tilde{y})$ in $\widetilde{\mathcal{Y}}$. Thus, it cannot be applied to signals $\tilde{w} \notin \widetilde{\mathcal{Y}}$ that cannot be produced as output from the model under permissible inputs $u \in \mathcal{U}$. Since the function space $\widetilde{\mathcal{Y}}$ was defined to exclusively contain the possible output signals of the model $\widetilde{\Sigma}$, it is obvious that the right inverse exists for each trajectory $\tilde{y} \in \widetilde{\mathcal{Y}}$. This holds true since $\widetilde{\mathcal{Y}}$ was defined using the shape of all possible outputs $\tilde{y}$ rather than their norm.

Remark 3.7 ([80]). Let $\widetilde{\Sigma}$ be a linear system and let $\widetilde{\Sigma}(s)$ denote its Laplace-transform. Then, its right inverse is given by $\widetilde{\Sigma}^{\mathrm{r}} = \widetilde{\Sigma}^{-1}(s)$, if $\widetilde{\Sigma}^{-1}(s)$ exists.

Remark 3.8 (Left Inverse). Although the left inverse does not play any role in this thesis, its difference to the right inverse is noted for the sake of completeness: The left inverse $\widetilde{\Sigma}^{\mathrm{l}}$ is defined by $\widetilde{\Sigma}^{\mathrm{l}}\widetilde{\Sigma}u = u$ and implies injectivity of the I/O map $\widetilde{\Sigma}$ while the right inverse implies surjectivity

> [80]. The left inverse *reconstructs* the *specific* input signal u that has been used to obtain a *measured* model output $\tilde{y}$. Thus, the left inverse does not exist if two different input signals $u_1 \neq u_2$ lead to the same output, i. e., $\tilde{y} = \widetilde{\Sigma} u_1 = \widetilde{\Sigma} u_2$. In such a case, a left inverse does not exist since it is impossible to detect which input was used to obtain the sensed output $\tilde{y}$. However, in this case the right inverse does exist since it would only have to select any one of the two possible input signals u_1 or u_2 which yields the requested output $\tilde{y} = \tilde{y}_{\mathrm{d}}$.

Right invertibility. The property of right invertibility is introduced below and deserves some words of introduction. In [80], it is established that right invertibility directly relates to surjectivity of the I/O map $\widetilde{\Sigma}$. Here, the output function space $\widetilde{\mathcal{Y}}$ (cf. Eq. (3.3)) was chosen such that the I/O map $\widetilde{\Sigma}$ is surjective. Thus, right invertibility would follow for the introduced plant model (3.2). However, with this choice of the output function space $\widetilde{\mathcal{Y}}$, a right inverse exists for all systems (3.2) and includes such systems that are only able to produce a unique output trajectory $\tilde{y}$. Such systems essentially behave like autonomous systems. Consider the following example.

> **Example 3.1:**
> Consider the system $\dot{x} = u$ with an output $\tilde{y} = 1$ which is uninfluenced by the states and some initial condition $\boldsymbol{x}(0) = x_0$. Clearly, there is only one possible output function, namely $\tilde{y} = 1$, $\forall t$, regardless of the choice of the input u. The system output behaves autonomously. Therefore, the set of output trajectories $\widetilde{\mathcal{Y}}$ only contains the single signal $\tilde{y} = 1$. Hence, inversion does not make any sense since the system can only generate this single signal. ■

Similarly, a system only has a single output trajectory $\tilde{y}$ if the set of input signals $\mathcal{U}$ only contains a single function u. In summary, inversion of systems with an autonomous I/O behaviour is senseless as nothing but their autonomous behaviour can be achieved.

Thus, in this thesis, the property of right invertibility will be understood as the feasibility of inversion. In other words, a system (3.2) will be considered right invertible if it makes sense to design a controller based on an inverse. This is clearly not the case if the system can only produce a unique output signal. Hence, for a given initial condition $\boldsymbol{x}_0$, the output-function space $\widetilde{\mathcal{Y}}$ must contain more than one signal for inversion to be useful.

Definition 3.5 (Right Invertibility). A system 3.2 is considered right invertible, if the set of its output function space $\widetilde{\mathcal{Y}}$ contains more than one signal. ◇

Remark 3.9 (Interpretation of right invertibility). The property of right invertibility is defined differently than right invertibility as discussed, for example in [80]. In [80], the input function space $\mathcal{U}$ as well as the output function space $\widetilde{\mathcal{Y}}$ were defined as sets that contain all analytic functions. Thus, they were defined less restrictive than in this thesis. In that case, the question about invertibility of a plant appears differently. The question becomes as to what property a dynamical system $\widetilde{\Sigma}$ should possess such that it can produce any arbitrary (but analytic) output function $\tilde{y}$. Or, worded differently, what property must a dynamical system possess such that it is surjective if its domain and its range consists of all analytic functions? This, however, is not conducive to an inversion-based control design since then the introduction of input constraints would suffice to render virtually any system non-invertible.

In this thesis the property of invertibility does not have the same importance or meaning since surjectivity of the I/O map is established by selecting a model-dependent output function space $\widetilde{\mathcal{Y}}$ (see Proposition 3.2). Since the problem of right inversion is approached by defining the output function space $\widetilde{\mathcal{Y}}$ such that the I/O map $\widetilde{\Sigma}$ is surjective on the input function space $\mathcal{U}$, a right inverse always exists. However, the set of signals on which the right inverse may be applied to, may, in the worst case, only consist of a single signal. An inversion of such a model is to be avoided.

With the introduction of the right inverse, the term (non-) minimum-phase can be defined [60, 61].

Definition 3.6 (Minimum-phase behaviour). The model $\widetilde{\Sigma}$ is said to be minimum phase if its right inverse $\widetilde{\Sigma}^{\mathrm{r}}$ is finite-gain $\mathcal{L}_p$-stable. Otherwise, the model $\widetilde{\Sigma}$ is said to be non-minimum phase. ◇

Remark 3.10 (Minimum-phase behaviour). In the definition above, non-minimum phase behaviour is given as it is most helpful for this thesis. Note, however, that the term non-minimum phase was originally defined for linear systems with right-half plane zeros or time-delays. Hence, linear systems are non-minimum phase if their phase cannot be deduced by their amplitude-over-frequency behaviour alone. Thus, for linear systems, the property of minimum phase and non-minimum phase behaviour is defined by a property of its transfer function which is an I/O operator. Definition 3.6 coincides with linear system theory in such that (in-) stability of the system inverse directly results from (non-) minimum phase behaviour of the system.

For nonlinear systems, it has then been defined by [51] as the instability of the zero dynamics. Interestingly, this definition uses a property of the system states (and not directly its I/O behaviour). As this thesis focuses on I/O behaviour instead of the behaviour of some states it is necessary to describe this property as a property of the I/O behaviour of nonlinear systems.

Note that the notion of zero dynamics [51] has not yet been introduced and, thus, (non-) minimum phase behaviour is not defined as (in)stability of the zero dynamics. In Section 5.3 non-minimum phase behaviour is discussed in more detail and the notion of internal dynamics and its stability is then used to define non-minimum phase behaviour. In this chapter, however, a system is defined by its I/O behaviour and the definition above has been chosen to accommodate this fact.

3.4 *Structure and Properties of Nonlinear SISO IMC*

This section gives a survey of the IMC structure and shows that all properties of the linear case (see Section 2.1.3) also hold in the nonlinear case with the necessary change in notation.

IMC structure. Figure 3.3 shows the IMC structure with nonlinear

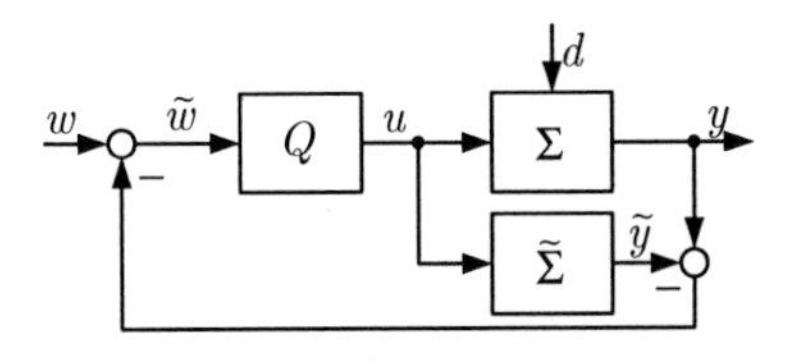

Fig. 3.3: IMC Structure for nonlinear plants.

IMC controller Q, nonlinear plant Σ and nonlinear plant model $\tilde{\Sigma}$. Disturbances d on the plant Σ include input and output disturbances as well as internal disturbances which affect the model error. The block diagram is identical to the classical IMC as shown in Fig. 2.1 with the exception of how the disturbances d enter the control structure.

Also, all observations obtained for the classical IMC given in Chapter 2 concerning the function of IMC still hold. This can be observed since the Figures 2.2 and 2.3 can also be obtained in the nonlinear case from which one concludes (as in the linear case) the following:

- A nonlinear IMC controller Q can be interpreted as a feedforward controller for the model $\widetilde{\Sigma}$.
- A nonlinear IMC controller Q is used in feedback to attenuate disturbances d and model uncertainties $\Sigma - \widetilde{\Sigma}$.

The following shows that all IMC properties as given in Section 2.1.3 also hold in the nonlinear case with the necessary change in notation.

IMC properties. The block diagram in Fig. 3.3 yields

$$\tilde{w} = w - y + \tilde{y} \tag{3.23}$$

$$\tilde{y} = \widetilde{\Sigma}\, Q\, \tilde{w}. \tag{3.24}$$

From Eqns. (3.23) and (3.24) the following properties can be derived [28].

Property 3.1 (Nominal Stability). *Assume an exact model ($\widetilde{\Sigma} = \Sigma$) in the absence of disturbances ($d = 0$). Then, the closed-loop system in Fig. 3.3 is internally stable if the controller Q and the plant $\widetilde{\Sigma}$ are stable.*

Proof. In the case of an exact model ($\widetilde{\Sigma} = \Sigma$) and no disturbances ($d = 0$), the IMC structure degenerates to the feedforward control shown in Fig. 2.2, which is internally stable for the conditions mentioned above. □

Property 3.2 (Perfect Control). *Assume that the right inverse of the model $\widetilde{\Sigma}^{\mathrm{r}}$ exists and that the closed-loop system is stable with controller $Q = \widetilde{\Sigma}^{\mathrm{r}}$. Then, the control will be perfect ($y = w$) for arbitrary disturbances d.*

Proof. Substituting Eq. (3.24) with $Q = \widetilde{\Sigma}^{\mathrm{r}}$ into Eq. (3.23) yields $w = y$. □

Remark 3.11 (Perfect Control). It is very important to realise that the property of perfect control *does not* necessitate an exact model, *nor* does it necessitate the absence of disturbances. In fact, when the IMC loop is stable, the above property states that perfect control is achieved even if the model has never been designed to resemble the plant, but rather was chosen arbitrarily. Also note that it is *only the model* which needs to be inverted. However, as will be discussed later on, an inverse alone is usually not realisable and thus, this property will remain an interesting but useless fact. The reader may go back to page 23 where this issue is briefly investigated for linear systems.

Property 3.3 (Zero Offset). *Assume that, for a steady-state signal* $\lim_{t\to\infty} \tilde{w} = \tilde{w}_{\mathrm{ss}}$*, the IMC controller* Q *acts as a right inverse in steady-state, i. e.,*

$$\lim_{t\to\infty} \left(\widetilde{\Sigma}\, Q\, \tilde{w}_{\mathrm{ss}}\right) = \tilde{w}_{\mathrm{ss}} \tag{3.25}$$

holds, and that the closed-loop system is stable. Then, offset-free control $y_{\mathrm{ss}} = w_{\mathrm{ss}}$ *is attained for asymptotically constant reference signals* $\lim_{t\to\infty} w = w_{\mathrm{ss}}$ *and constant disturbances* $\lim_{t\to\infty} d = d_{\mathrm{ss}}$.

Proof. With Eq. (3.25) inserted into Eq. (3.24) one gets

$$\tilde{y}_{\mathrm{ss}} = \tilde{w}_{\mathrm{ss}} \tag{3.26}$$

for $t \to \infty$ with $\lim_{t\to\infty} \tilde{w} = \tilde{w}_{\mathrm{ss}}$. Substituting Eq. (3.26) into Eq. (3.23) (and taking the limit as $t \to \infty$) yields

$$y_{\mathrm{ss}} = w_{\mathrm{ss}}.$$

□

Property 3.4 (Robust Stability). *Assume that* $\widetilde{\Sigma}$ *and* Q *are stable systems, for which the IMC structure in Fig. 3.3 is stable for an exact model* $\widetilde{\Sigma} = \Sigma$*. Then, the IMC structure remains stable if the model deviation* $\Sigma - \widetilde{\Sigma}$ *satisfies the gain inequality (see Definition 3.2)*

$$g\left((\widetilde{\Sigma} - \Sigma)Q\right) < 1. \tag{3.27}$$

The proof follows directly from the Small-Gain Theorem.

Comparing Properties 3.1 to 3.4 of the nonlinear IMC with the Properties 2.1 to 2.4 one finds that the structural properties of IMC remain the same, independently whether linear or nonlinear systems are to be controlled. Hence, just as in the linear case,

- the stability of the plant Σ and the controller Q are necessary conditions for closed-loop stability,
- the property of perfect control cannot be realised either as will be discussed in the following section, and
- the IMC structure guarantees zero steady-state offset and a certain robust stability.

From the above, one concludes that with the properties also all interpretations of the function of IMC that have been obtained in the linear case carry over to the nonlinear case with the necessary change in notation. Thus, the same design procedure applies [28].

3.5 IMC Design Procedure for Minimum Phase Models with Well-Defined Relative Degree

For the sake of a straightforward introduction of the concept of nonlinear IMC design, the system class is limited as follows: This section deals with finite-gain stable (cf. Definition 3.1), minimum-phase systems (cf. Definition 3.6) represented by the plant model (3.2). Moreover, the plant model should have a well-defined relative degree r and it is first assumed that the set $\mathcal{U}$ of permissible control signals contains all arbitrary but finite functions (i. e., $u \in \mathcal{U} = \mathcal{L}_\infty$).

The restriction to stable systems is inherent in the IMC structure (see Property 3.1) and invertibility of the plant model $\widetilde{\Sigma}$ is a necessary condition for IMC design. These two conditions are not removed throughout this thesis. The restriction to minimum-phase systems implies stable inverses, which allows for a simple controller design procedure since nominal stability is guaranteed (cf. Property 3.1). The restriction to a well-defined relative degree r means that Proposition 3.2 applies. These properties will be exploited below.

The assumptions of a minimum-phase model and a well-defined relative degree r severely limit the system class and exclude many automotive plants including the two-state turbocharged diesel engine which is discussed in Chapter 7. These excessive assumptions are dropped in Chapter 5, where the basic idea of IMC design is extended to stable systems with an ill-defined relative degree r and to non-minimum phase systems. In that same chapter, the permissible set of control signals $\mathcal{U}$ will be restricted to incorporate input constraints.

3.5.1 Control Goal and IMC Design Procedure

Control goal. The goal is to find an IMC controller Q such that the following requirements are satisfied:

- The output y of the nonlinear plant Σ tracks an arbitrary but finite reference signal $w \in \mathcal{W} = \mathcal{L}_\infty$ with zero steady-state offset (assuming a constant steady-state value, i. e., $y_{\mathrm{ss}} = w_{\mathrm{ss}}$ for $w_{\mathrm{ss}} = \lim_{t\to\infty} w = \text{const}$ and $d_{\mathrm{ss}} = \lim_{t\to\infty} d = \text{const}$).
- With t as the current time, future reference values $w(\tau)$ with $\tau > t$ are unknown and the controller can react only in dependence on the current and earlier values. Thus, a non-predictive control scheme is desired.

Main idea. The main idea is to construct the nonlinear IMC controller Q as the composition of a *linear* filter F and the right inverse $\widetilde{\Sigma}^{\mathrm{r}}$ of the model $\widetilde{\Sigma}$, namely

$$\boxed{Q = \widetilde{\Sigma}^{\mathrm{r}}\, F}\,. \tag{3.28}$$

The resulting IMC structure is shown in Fig. 3.4. Figure 3.4 differs from

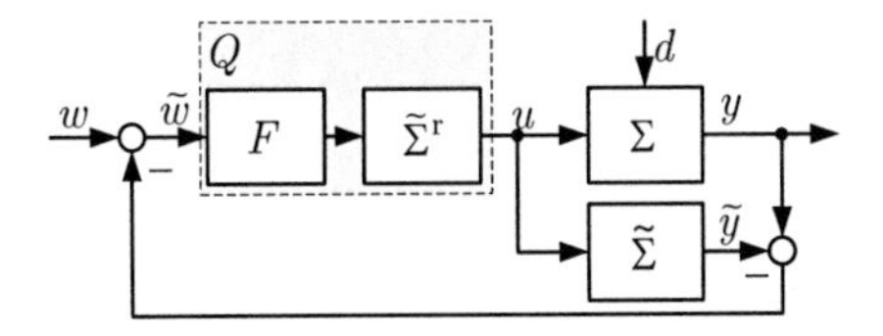

Fig. 3.4: Generalised IMC Structure.

Fig. 2.7 in such that, in the linear case, the IMC controller Q is represented as a single transfer function for implementation. Here, however, the filter F and the inverse $\widetilde{\Sigma}^{\mathrm{r}}$ are each implemented individually, as shown by the dashed and solid lines.

Although this approach follows the design procedure for linear systems (see Eq. (2.14a)), it needs to be explained differently.

Realisability of the right inverse. According to Property 3.2, the desired IMC controller is the right inverse $Q = \widetilde{\Sigma}^{\mathrm{r}}$ of the model $\widetilde{\Sigma}$ because perfect control follows. However, this controller is not realisable: As introduced in Section 3.3, the right inverse $\widetilde{\Sigma}^{\mathrm{r}}$ maps signals from the output-function space of the model $\widetilde{\mathcal{Y}}$ into the range $\mathcal{U}$ of the permissible input signals. In the IMC structure (Fig. 3.3), the signal $\tilde{w} \in \widetilde{\mathcal{W}}$ is the input signal for Q. Since the reference signal may be arbitrary but limited (i. e., $w \in \mathcal{L}_\infty$) one concludes that also the filter input $\tilde{w}$ lies in the same space ($\tilde{w} \in \mathcal{L}_\infty$). Hence, for the input $\tilde{w}$ of the controller Q, the relationship

$$\tilde{w} \in \mathcal{L}_\infty \supset C^{r-1} \supseteq \widetilde{\mathcal{Y}} \quad \text{for } r > 0 \tag{3.29}$$

holds. That is, the signal $\tilde{w}$ is not necessarily in the domain of the right inverse $\widetilde{\Sigma}^{\mathrm{r}}$ and, hence, $\widetilde{\Sigma}^{\mathrm{r}}$ cannot be used as controller Q. Therefore, a pure right inverse as IMC controller is not realisable.

Remark 3.12. For $r = 0$, the IMC controller as well as the model have a direct feedthrough and the IMC structure results in an algebraic loop which is an ill-posed feedback loop [83] and, thus, cannot be realised.

Function of the IMC filter. To circumvent this difficulty, the IMC filter F is introduced as an operator which fulfils

$$F : \mathcal{L}_\infty \to C^{r-1}.$$

It produces for every $\tilde{w} \in \mathcal{L}_\infty$ an output $\tilde{y}_\mathrm{d} \in \widetilde{\mathcal{Y}}$ that is acceptable as input to the right inverse $\widetilde{\Sigma}^{\mathrm{r}}$. The interpretation of this idea is that the IMC filter acts as a translator from the signal $\tilde{w} \in \widetilde{\mathcal{W}}$ into the set $\widetilde{\mathcal{Y}}$ of signals that the right inverse "understands." The block diagram in Fig. 3.5 clarifies the above change in signals through the feedforward path of the IMC structure.

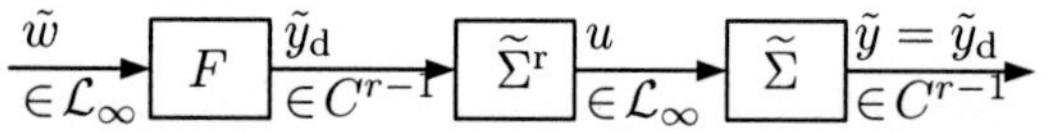

$$\xrightarrow[\in \mathcal{L}_\infty]{\tilde{w}} \boxed{F} \xrightarrow[\in C^{r-1}]{\tilde{y}_\mathrm{d}} \boxed{\widetilde{\Sigma}^{\mathrm{r}}} \xrightarrow[\in \mathcal{L}_\infty]{u} \boxed{\widetilde{\Sigma}} \xrightarrow[\in C^{r-1}]{\tilde{y} = \tilde{y}_\mathrm{d}}$$

Fig. 3.5: Signal spaces from IMC filter input $\tilde{w}$ to model output $\tilde{y}$.

As in the linear case, the IMC controller Q given in Eq. (3.28) can be interpreted as a feedforward controller for the model $\widetilde{\Sigma}$ as can be seen from the equalities

$$\begin{aligned} \tilde{y} &= \widetilde{\Sigma} Q \tilde{w} = \widetilde{\Sigma}\,(\widetilde{\Sigma}^{\mathrm{r}} F)\,\tilde{w} = (\widetilde{\Sigma}\,\widetilde{\Sigma}^{\mathrm{r}}) F\,\tilde{w} \\ &= F\tilde{w} = \tilde{y}_\mathrm{d}. \end{aligned} \tag{3.30}$$

If zero steady-state offset is desired (Property 3.3), the IMC filter F has to fulfil the condition

$$\lim_{t\to\infty} F\tilde{w}_{\mathrm{ss}} = \tilde{w}_{\mathrm{ss}}. \tag{3.31}$$

This property is denoted by $F_{\mathrm{ss}} = 1$ in the following and means the equivalent of a unitary steady-state gain. Note that an ill-posed feedback loop is avoided if F has a relative degree of $r > 0$.

The proposed IMC design procedure is summarised in the following algorithm:

Algorithm 3.1. ***Generalised IMC Design Procedure***

Given: A stable and invertible minimum phase plant model $\widetilde{\Sigma}$

Step 1: Compute the model right inverse $\widetilde{\Sigma}^{\mathrm{r}}$

Step 2: Find an IMC filter $F : \widetilde{\mathcal{W}} \to \widetilde{\mathcal{Y}}$ *which*

- *has a "good" step response from* $\tilde{w}$ *to* $\tilde{y}$*, and*
- *has a steady-state gain of one.*

Step 3: With $Q = \widetilde{\Sigma}^{\mathrm{r}} F$*, the IMC control loop is given in Fig. 3.4.*

Result: Nonlinear output feedback IMC control loop.

Note that the derivation of the right inverse (Step 1) is given in Chapter 4 for flat and input affine SISO systems. In the following, the focus lies on the IMC filter F and its composition with the right inverse $\widetilde{\Sigma}^{\mathrm{r}}$.

Composition of the right inverse and IMC filter. In the linear case, the implementation of the series of $F(s)$ and $\widetilde{\Sigma}(s)$ is done by multiplying the two transfer functions to obtain $Q(s) = \widetilde{\Sigma}(s)F(s)$ and, as Q is proper, one can implement the complete IMC controller $Q(s)$ as one transfer function. In the nonlinear case, however, the implementation of the IMC controller is mathematically more involved.

The domain and range of any (nonlinear) operator are signal spaces (cf. Section 3.1). These signals (e. g., $\tilde{y}$) are defined over all time (i. e., $t \in (-\infty, \infty)$). Hence, it is possible to exactly compute all existing derivatives of these signals at any instance in time τ as the signal value for $t > \tau$ is known. Therefore, by interpreting the right inverse $\widetilde{\Sigma}^{\mathrm{r}}$ of the model $\widetilde{\Sigma}$ as a SISO operator (Fig. 3.2), there is no problem using any number of derivatives $\tilde{y}_{\mathrm{d}}^{(i)}$ (where $i = 1, \ldots, r$) of the input signal $\tilde{y}_{\mathrm{d}}$ [48, 80] to compute the output u:

$$\begin{aligned} u &= \widetilde{\Sigma}^{\mathrm{r}} \tilde{y}_{\mathrm{d}} \\ u(t) &\triangleq \widetilde{\Sigma}^{\mathrm{r}} \left(\tilde{y}_{\mathrm{d}}(t), \dot{\tilde{y}}_{\mathrm{d}}(t), \ldots, \tilde{y}_{\mathrm{d}}^{(r)}(t) \right), \end{aligned} \tag{3.32}$$

where t is the current time.

However, implementation of the right inverse $\widetilde{\Sigma}^{\mathrm{r}}$ means to provide $\widetilde{\Sigma}^{\mathrm{r}}$ with all necessary information, at any instance in time, so that it can compute the output u. This means that the derivatives $\tilde{y}_{\mathrm{d}}^{(i)}$ of the input $\tilde{y}$ must be provided to $\widetilde{\Sigma}^{\mathrm{r}}$ for the following reason: The control problem concerns reference signals w, which are only known up to the current time t. Consequently, the input $\tilde{w}$ to the IMC filter and its output $\tilde{y}_{\mathrm{d}}$ are also defined only up to the current time t. Therefore, $\widetilde{\Sigma}^{\mathrm{r}}$ represents a non-realisable operator since the derivatives $\tilde{y}_{\mathrm{d}}^{(i)}(t)$ cannot be computed without the knowledge of future values of the model output $\tilde{y}(t)$.

Hence, it is proposed to *implement* the right inverse $\widetilde{\Sigma}^{\mathrm{r}}$ such that, in addition to the signal $\tilde{y}_{\mathrm{d}}$, it also takes the information of the derivatives $\tilde{y}_{\mathrm{d}}^{(j)}(t)$ with $j = 1, \ldots, r$ from the filter F (Fig. 3.6). The chain connection

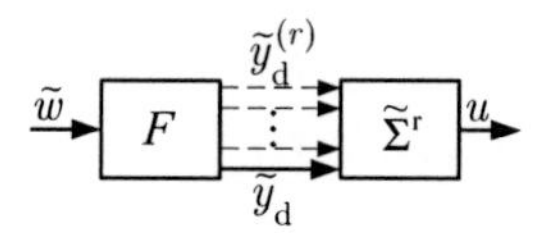

Fig. 3.6: I/O signals of the implementation of the composition of F and $\widetilde{\Sigma}^{\mathrm{r}}$.

of F and $\widetilde{\Sigma}^{\mathrm{r}}$ as shown in Fig. 3.6 forms a realisable (non-anticipatory) system Q. This implementation of $\widetilde{\Sigma}^{\mathrm{r}}$ is realisable for unknown future values of $\tilde{w}(\tau)$ with $\tau > t$ which means that the IMC controller Q is non-anticipative and, hence, technically realisable.

Implementation of the IMC filter. The IMC filter F maps its input signals $\tilde{w} \in \widetilde{\mathcal{W}}$ into the domain of the right inverse $\widetilde{\mathcal{Y}}$ (see Algorithm 3.1)

$$\tilde{y}_{\mathrm{d}} = F\tilde{w}. \tag{3.33}$$

The filter output $\tilde{y}_{\mathrm{d}}$ should follow its input signal $\tilde{w}$ "closely", and have a steady-state gain of $F_{\mathrm{ss}} = 1$. The domain $\widetilde{\mathcal{W}}$ is assumed to contain all limited functions (i. e., $\tilde{w} \in \mathcal{L}_\infty$).

It is proposed to use a *linear* IMC filter F with transfer function

$$F(s) = \frac{\tilde{y}_{\mathrm{d}}(s)}{\tilde{w}(s)} = \frac{1}{k_r s^r + k_{r-1} s^{r-1} + \cdots + 1} \tag{3.34}$$

and it is proposed to choose the k_i such that

$$F(s) = \frac{1}{\left(\frac{s}{\lambda} + 1\right)^r} \tag{3.35}$$

holds where λ can be chosen by the designer. The filter F has the relative degree r and, thus, (cf. Proposition 3.2) its output $\tilde{y}_{\mathrm{d}}$ is r-times differentiable ($\tilde{y}_{\mathrm{d}} \in C^{r-1}$). Hence, the filter F maps arbitrary limited signals $\tilde{w}$ into signals $\tilde{y}_{\mathrm{d}}$ which are in the function space $\tilde{y}_{\mathrm{d}} = \tilde{y} \in C^{r-1}$. Additionally, Eq. (3.31) holds with this IMC filter, since $F_{\mathrm{ss}} = 1$.

Remark 3.13 (Nonlinear IMC filters). In some publications on inversion based control (e. g., [2, 28]), it is mentioned that the use of nonlinear filters (or nonlinear reference systems) would result in better performance or robustness. Although nonlinear IMC filters will also be proposed later on for special cases, here, essentially a linear IMC filter is used since it offers a simple structure and is straightforward to tune.

Besides the output signal $\tilde{y}_{\mathrm{d}}$, the filter has to deliver the first r derivatives of $\tilde{y}_{\mathrm{d}}$. To this end, it is further proposed to implement F as a state-variable filter [97] as shown in Fig. 3.7. Such an implementation au-

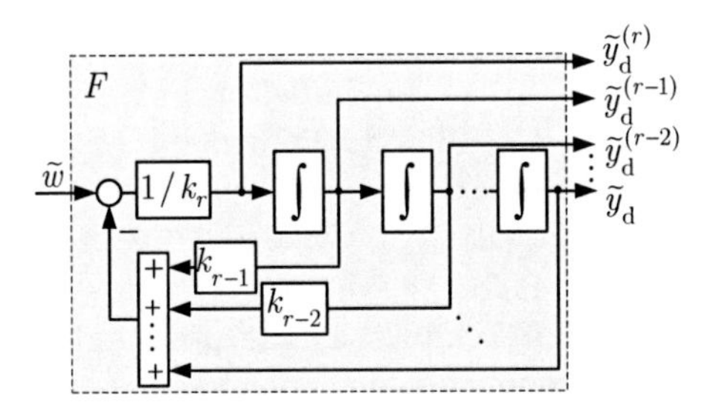

Fig. 3.7: IMC filter F implemented as an SVF.

tomatically delivers r derivatives of the output $\tilde{y}_{\mathrm{d}}$. The initial states of the integrators of the IMC filter F follow from Eq. (3.13). It only differs from the classical IMC filter in the fact that it also gives r derivatives of the output. The implementation of the composition of IMC filter F and model inverse $\widetilde{\Sigma}^{\mathrm{r}}$ results in the connection signals displayed in Fig. 3.6.

Note that the filter $F(s)$ with the transfer function (3.34) can also be implemented in state-space using the control normal form. In state-space, the filter F consists of the r-state vector $\boldsymbol{x}_{\mathrm{F}}(t) \in \mathbb{R}^r$, input $\tilde{w}$ and the $r+1$ outputs $[\tilde{y}_{\mathrm{d}}, \dot{\tilde{y}}_{\mathrm{d}}, \ldots, \tilde{y}_{\mathrm{d}}^{(r)}]^T$:

$$\dot{\boldsymbol{x}}_{\mathrm{F}} = \begin{bmatrix} 0 & 1 & 0 & \cdots & 0 \\ 0 & 0 & 1 & \cdots & 0 \\ \vdots & & & \ddots & 0 \\ 0 & 0 & \cdots & 0 & 1 \\ -\frac{1}{k_r} & -\frac{k_1}{k_r} & \cdots & & -\frac{k_{r-1}}{k_r} \end{bmatrix} \boldsymbol{x}_{\mathrm{F}} + \begin{bmatrix} 0 \\ 0 \\ \vdots \\ 0 \\ \frac{1}{k_r} \end{bmatrix} \tilde{w}$$

$$\begin{bmatrix} \tilde{y}_{\mathrm{d}} \\ \dot{\tilde{y}}_{\mathrm{d}} \\ \vdots \\ \tilde{y}_{\mathrm{d}}^{(r-1)} \\ \tilde{y}_{\mathrm{d}}^{(r)} \end{bmatrix} = \begin{bmatrix} 1 & 0 & 0 & \cdots & 0 \\ 0 & 1 & 0 & \cdots & 0 \\ 0 & 0 & \ddots & 0 & 0 \\ 0 & 0 & \cdots & 0 & 1 \\ -\frac{1}{k_r} & -\frac{k_1}{k_r} & \cdots & & -\frac{k_{r-1}}{k_r} \end{bmatrix} \boldsymbol{x}_{\mathrm{F}} + \begin{bmatrix} 0 \\ 0 \\ \vdots \\ 0 \\ \frac{1}{k_r} \end{bmatrix} \tilde{w} \tag{3.36a}$$

with the initial conditions from Eq. (3.13) as

$$\boldsymbol{x}_{\mathrm{F}}(0) = \left[h(\boldsymbol{x}_0), \quad L_f h(\boldsymbol{x}_0), \quad \cdots, \quad L_f^{r-1} h(\boldsymbol{x}_0)\right]^T. \tag{3.36b}$$

Interpretation of the IMC filter F**.** The IMC filter is chosen to have the same relative degree r as the plant model $\widetilde{\Sigma}$. Moreover, the initial condition (3.36b) of the IMC filter is chosen such that its output matches the initial shape (cf. Eq. (3.13)) of the model output $\tilde{y}$. Employing Proposition 3.1 and Proposition 3.2 one finds that the IMC filter F and the model $\widetilde{\Sigma}$ have the same output-function space $\widetilde{\mathcal{Y}}$. Therewith one finds that the inverse $\widetilde{\Sigma}^{\mathrm{r}}$ (cf. Definition 3.4) is always defined as it always operates on the output function space $\widetilde{\mathcal{Y}}$ of the model $\widetilde{\Sigma}$.

It is important to appreciate that the proposed *implementation* of IMC filter and right inverse shown in Fig. 3.6 does not change the fact that, from a functional analytic view, the right inverse $\widetilde{\Sigma}$ as well as the filter F still are SISO operators.

Remark 3.14. Note that $\tilde{y}_{\mathrm{d}}^{(r)}$ could be regarded as the only necessary information from the IMC filter F. This requires the integrator chain to be implemented once in the filter F and once in the inverse $\widetilde{\Sigma}^{\mathrm{r}}$ with identical initial conditions. This, however, is a waste of processing power and memory since the resulting IMC controller Q would have r redundant states. Therefore, the integrator chain is only implemented in the IMC filter.

Moreover, consider inverses of linear systems given in Laplace domain. The order of such an inverse is identical to the number of zeros of the model to be inverted. Clearly, the order of the inverse is neither the relative degree of the model to be inverted nor the number of its states. In this respect, the procedure proposed above represents an analogy to inversion of linear systems.

3.5.2 Substitution of the Internal Model

For a reduction of the implementation effort, the structure shown in Fig. 3.4 can be simplified.

Theorem 3.1. *Consider the IMC structure as shown in Fig. 3.4. The computation of the model output $\tilde{y}$ can be performed by $\tilde{y} = F\tilde{w}$. Therewith, the internal model $\widetilde{\Sigma}$ is redundant and can be substituted.*

If follows that the IMC structure in Fig. 3.4 is equivalent to the structure shown in Fig. 3.8, where no plant model appears explicitly.

The proof follows directly from Eq. (3.30). This theorem implies a dra-

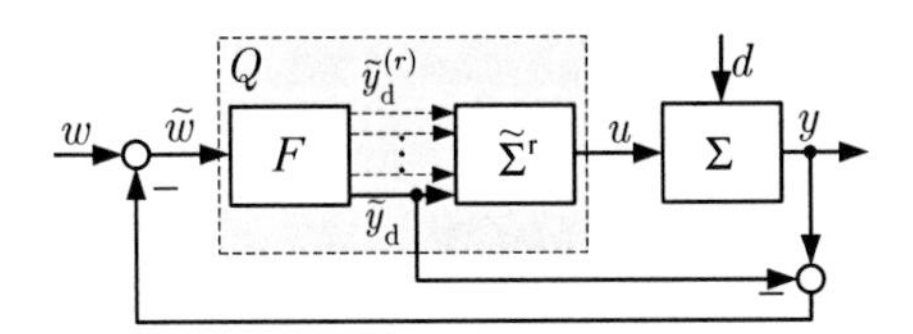

Fig. 3.8: Generalised IMC controller with no internal model.

matic complexity reduction of the final controller. In the original version shown in Fig. 3.4, the controller consists of the r-th order filter F, $(n-r)$-th order inverse $\widetilde{\Sigma}^{\mathrm{r}}$ and the n-th order plant model and has the overall order of $2n$, whereas in Fig. 3.8 the plant model is omitted, resulting in an overall order n.

This interesting result is achieved since the signal $\tilde{y}_{\mathrm{d}}$ between IMC filter F and right inverse $\widetilde{\Sigma}^{\mathrm{r}}$ is available in the final implementation (see Fig. 3.6). With this result, it is proposed to replace Step 3 of Algorithm 3.1 by:

Algorithm 3.2. ***IMC control loop implementation***
Algorithm 3.1 with the following replacement:
Step 3: *Compute the model output $\tilde{y}$ by $\tilde{y} = F\tilde{w}$ and omit the internal model (see Fig. 3.8).*

Interestingly, in the literature it has not yet been proposed to substitute the internal model in the IMC loop for linear systems. This is obvious for two reasons:

- In the linear case, the IMC controller $Q(s)$ is always implemented as a single transfer function. Thus, the output signal $\tilde{y}_{\mathrm{d}}$ of the filter $F(s)$ is not available.
- A minimal realisation of a linear IMC loop can be obtained by computing the equivalent controller $C(s)$ for a classical control loop from Eq. (2.3). Hence, there is another way to reduce the order of the final controller.

3.5.3 Robust Stability for Unstructured Uncertainties

Suppose that the plant model $\widetilde{\Sigma}$ has an unstructured multiplicative uncertainty Δ such that (see Fig. 3.9)

$$\Sigma = (I + \Delta)\widetilde{\Sigma} \tag{3.37}$$

holds. The uncertainty Δ can be interpreted as some nonlinear dynamical

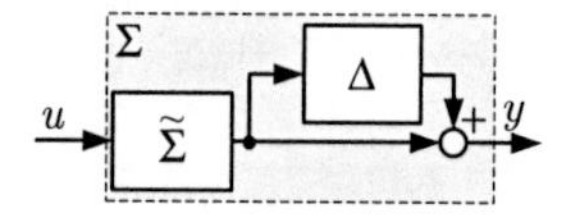

Fig. 3.9: Plant Σ represented with multiplicative output uncertainty Δ.

system. Then, with Eq. (3.22) and the nonlinear IMC structure from Fig. 3.8 without disturbances ($d = 0$) one finds as open loop system Σ_0 from filter input signal $\tilde{w}$ to feedback signal $y - \tilde{y}$

$$\begin{aligned} \Sigma_0 &= (\Sigma - \widetilde{\Sigma})\widetilde{\Sigma}^{\mathrm{r}} F = (\Sigma\widetilde{\Sigma}^{\mathrm{r}} - I)F \\ &= \Delta F \end{aligned} \tag{3.38}$$

which is shown in Fig. 3.10.

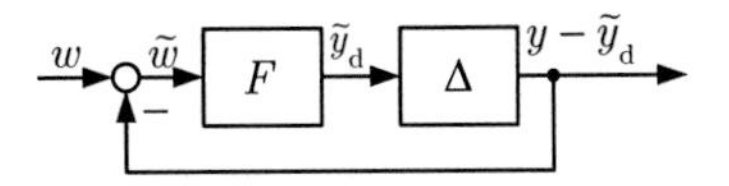

Fig. 3.10: Resulting IMC control loop with multiplicative output uncertainty Δ.

First, robust stability of the control structure in Fig. 3.10 is shown using the Small-Gain Theorem: Let $\bar{\Delta}$ be an upper boundary for the uncertainty Δ, such that

$$g(\Delta) \leq \bar{\Delta} \tag{3.39}$$

holds, where $g(\Delta)$ is the gain of the system operator Δ (cf. Definition 3.2). The Small-Gain Theorem [99] says that if Δ and F are stable systems and if the open-loop gain is smaller than one, i. e.,

$$g(\Delta F) < 1 \tag{3.40}$$

holds, then closing the loop yields a stable feedback system. Hence, the closed-loop is stable if

$$\bar{\Delta} \cdot g(F) < 1 \tag{3.41}$$

holds. Since the IMC filter F was designed as a linear system, well known methods from linear control theory can be employed to determine its gain:

$$g(F) = \|F(s)\|_\infty = \max_\omega |F(j\omega)|. \tag{3.42}$$

The H_∞-norm in frequency domain relates to the $\|\cdot\|_{\mathcal{L}_2}$ signal norm [54]. Thus, let $g(\Delta)$, and therewith $\bar{\Delta}$, be defined using the $\|\cdot\|_{\mathcal{L}_2}$ signal norm. The following robustness test follows directly from the Small-Gain Theorem:

Theorem 3.2 (Robust Stability of an IMC Loop). *An IMC loop is robustly stable for the nonlinear plant given in Eq.* (3.37) *with the upper uncertainty boundary* $\bar{\Delta}$ *if*

$$\bar{\Delta} \cdot \|F(s)\|_\infty < 1 \tag{3.43}$$

holds.

As the filter parameters can be chosen arbitrarily and

$$\|F(s)\|_\infty = \max_\omega |F(j\omega)| = |F(0)| = 1,$$

Eq. (3.43) shows a considerable robustness of the nonlinear IMC loop, because stability is ensured for $\bar{\Delta} < 1$.

The above shows a structural robustness of the IMC feedback loop. Therewith, a nonlinear IMC control has an advantage over some other approaches, such as exact linearisation, for which no conceptual robustness has been shown. However, the robustness boundaries are conservative since they are only dependent on the gains of F and Δ and do not account for their actual behaviour.

A less conservative robustness test can be given if one further exploits the fact that the feedback loop shown in Fig. 3.10 consists of a linear operator, namely F, and a nonlinear operator, namely Δ. For such feedback loops, specialised robustness criteria can be given. The reader is referred to the literature cited in [54, 99] for an overview. Here, it is proposed to employ the Circle Theorem as developed in [99]. The Circle Theorem employs a Nyquist-like stability criterion where the location of the Nyquist curve of $F(s)$ defines the boundaries of the uncertainty Δ as conic regions in the instantaneous[3] I/O behaviour. Although the Circle Theorem is also not a necessary condition for stability, it is significantly less conservative than the Small-Gain Theorem. A thorough review of the Circle Theorem is beyond the scope of this thesis and since it is directly applicable to the feedback structure of Fig. 3.10 and its introduction is not necessary to follow the nonlinear IMC design, the reader is referred to [99].

3.5.4 Application to Linear Plants

In this section, the previous results are briefly compared to the classical IMC design method for linear systems as discussed in Chapter 2.

Theorem 3.3 (Analogy of the IMC design for nonlinear systems to the classical IMC design). *For linear systems $\widetilde{\Sigma}(s)$, the proposed IMC design for nonlinear systems using the design law* (3.28) *and the IMC filter from Eq.* (3.34) *with the proposed implementation as shown in Fig. 3.6 yields the same IMC controller Q as the classical design law* (2.14a) *with classical IMC filter from Eq.* (2.14b).

Proof. According to Remark 3.7, the (right) inverse is given by $\widetilde{\Sigma}^{\mathrm{r}} = \widetilde{\Sigma}(s)^{-1}$. Therewith, both the IMC design law for linear plants (cf.

[3] Instantaneous in the sense that only signal values of the input and output at the current time t are regarded.

Eq. (2.14a)) and the proposed IMC design law for nonlinear plants (cf. Eq. (3.28)) come to the same result, namely

$$Q(s) = \widetilde{\Sigma}(s)^{-1} F(s). \tag{3.44}$$

The inverse $\widetilde{\Sigma}^{-1}(s)$ is

$$\widetilde{\Sigma}^{-1}(s) = \frac{u(s)}{\tilde{y}_{\mathrm{d}}(s)} = \frac{a_0 + a_1 s + \ldots + a_{n-1} s^{n-1} + s^n}{b_0 + b_1 s + \ldots + b_q s^q}, \tag{3.45}$$

which is an improper transfer function. It can be rewritten as

$$\begin{aligned} u(s) = \; & \overbrace{\frac{1}{b_0 + b_1 s + \ldots + b_q s^q}}^{\text{Part 1}} \cdot \\ & \overbrace{\left(a_0 \left[\tilde{y}_{\mathrm{d}}(s)\right] + a_1 \left[s\tilde{y}_{\mathrm{d}}(s)\right] + \ldots + a_r \left[s^r \tilde{y}_{\mathrm{d}}(s)\right]\right) +}^{\text{Part 2}} \\ & \underbrace{\frac{a_{r+1} s + \cdots + a_n s^q}{b_0 + b_1 s + \ldots + b_q s^q} \cdot \left[s^r \tilde{y}_{\mathrm{d}}(s)\right]}_{\text{Part 3}}. \end{aligned} \tag{3.46}$$

The *available* outputs of the SVF F are in their Laplace-transform $\tilde{y}_{\mathrm{d}}(s)$, $s\tilde{y}_{\mathrm{d}}(s), \ldots, s^r \tilde{y}_{\mathrm{d}}(s)$. Every part of Eq. (3.46) is by itself realisable (i. e., its value can be computed at time t) by inserting the output of F: Part 1 of Eq. (3.46) is strictly proper, Part 2 of Eq. (3.46) is a linear combination of the filter outputs $\left[s^j \tilde{y}_{\mathrm{d}}(s)\right]$ $(j = 0, \ldots, r)$. Finally, Part 3 of (3.46) is a proper transfer function multiplied by the last filter output.

Thus, the linear IMC controller $Q(s)$ can be implemented as it is proposed for the nonlinear IMC controller Q, namely with the connection signals shown in Fig. 3.6. □

Thus, even with a linear IMC controller $Q(s)$, the IMC filter $F(s)$ and model inverse $\widetilde{\Sigma}^{-1}(s)$ can be implemented independently of each other using the connection signals displayed in Fig. 3.6 despite of $\widetilde{\Sigma}^{-1}(s)$ being improper. Hence, the proposed nonlinear IMC design is a *direct* extension of the classical IMC design to nonlinear systems. In this respect, this proposed nonlinear IMC design method is unique since other extensions of IMC to the nonlinear case do not yield the same IMC controller *structure* for linear systems.

3.6 Feasibility of Nonlinear IMC as Automotive Controller

In Section 2.3, the linear IMC controller was evaluated and it was found that a linear IMC is a feasible automotive controller since it fulfils the demands listed in Section 1.3.2. The proposed design of a nonlinear IMC is a direct extension of the linear IMC design and all structural properties of the IMC structure hold true for the nonlinear case. Section 2.3 explains that a linear IMC controller is a feasible automotive controller. This conclusion is based on the design and the structural properties of IMC and, hence, completely carries over to the nonlinear case. Thus, one concludes that the proposed nonlinear IMC controller is a feasible automotive controller.

3.7 Summary

In this chapter, the IMC controller Q was introduced as the series of the IMC filter F and the right inverse $\widetilde{\Sigma}^{\mathrm{r}}$ of the model $\widetilde{\Sigma}$. The requirement for this method is a stable and minimum-phase model $\widetilde{\Sigma}$. It was shown that, for nonlinear systems, such an IMC controller yields a nominally stable closed loop which produces zero steady-state offset. If the plant Σ can be represented by a model with output uncertainties, a closed-loop robust stability analysis is possible if an upper boundary of the gain of the output uncertainty is given. A considerable robustness concerning stability can be expected.

The internal model, which is an important part of the internal model control method, can generally be omitted since its output $\tilde{y}$ is equal to the output of the IMC filter F (i. e., $\tilde{y}_{\mathrm{d}} = \tilde{y}$). Although omitting the internal model does not offer new properties of the closed loop, it will reduce the order of the overall controller to be implemented.

However, a method to obtain a right inverse has not been discussed. This will be done in the following for the system classes of flat and input-affine systems. Once a right inverse of any model is established, it can be implemented such that the proposed IMC filter can be used, resulting in a robustly stable nonlinear output feedback control loop.

Note that not before Chapter 5 will the demands on a well-defined relative degree and a minimum-phase model be dropped.

4. INTERNAL MODEL CONTROL DESIGN FOR INPUT-AFFINE SISO SYSTEMS AND FLAT SISO SYSTEMS

In the preceding chapter, the concept of nonlinear IMC design has been introduced by considering mainly the I/O behaviour of dynamical systems. A state-space description of the IMC loop could be avoided since all necessary properties of the closed-loop have been shown using I/O considerations alone.

The right inverse has been defined by the direction of the mapping between the function spaces $\widetilde{\mathcal{Y}}$ and $\mathcal{U}$. This chapter addresses the problem of finding a right inverse for a given dynamical system. The system class for which this is discussed includes flat systems and systems in I/O normal form. For this purpose, however, a state-space approach needs to be taken as a tool to create a right inverse which can be used in the IMC design. Note that the state-space considerations will be limited to the construction of the IMC controller and do not concern the closed-loop behaviour. Note that the closed-loop behaviour has been explained in the previous chapter using I/O considerations alone.

This chapter is divided into Section 4.1, which addresses inverses for differentially flat systems, and Section 4.2, which addresses inverses for input-affine systems. Those two sections are independent of each other and the reader may skip either one.

4.1 IMC of Flat SISO Systems

This section reviews some basic properties of flat systems and employs these properties to construct the right inverse of the plant model $\widetilde{\Sigma}$ (Step 1 of the design Algorithm 3.2). The right inverse is then used to propose an IMC controller following the design idea presented in the previous chapter.

Differential flatness of dynamical systems. Consider a nonlinear SISO system $\widetilde{\Sigma}$ in state-space representation (3.2).

Definition 4.1 (Flatness [30, 81]). The system $\widetilde{\Sigma}$ is called *flat* if there is a variable $z(t)$ (called the *flat output*), such that the following conditions are satisfied:

1. The flat output $z(t)$ can be represented in terms of the state $\boldsymbol{x}(t)$

$$z(t) = \Phi\left(\boldsymbol{x}(t)\right). \tag{4.1}$$

2. The state $\boldsymbol{x}(t)$, the input $u(t)$, and their time derivatives can be represented in terms of $z(t)$ and a finite number of its time derivatives $\dot{z}, \ldots, z^{(n)}$:

$$\boldsymbol{x}(t) = \boldsymbol{\psi}_{\mathrm{x}}\left(z(t), \dot{z}(t), \ldots, z^{(n-1)}(t)\right) \tag{4.2a}$$

$$u(t) = \psi_{\mathrm{u}}\left(z(t), \dot{z}(t), \ldots, z^{(n)}(t)\right). \tag{4.2b}$$

◇

If the conditions in Eqns. (4.1) and (4.2) are satisfied then the output $\tilde{y}(t)$ can be expressed as [40]

$$\tilde{y}(t) = h\left(\boldsymbol{\psi}_{\mathrm{x}}\left(z(t), \ldots, z^{(n-1)}(t)\right), \psi_{\mathrm{u}}\left(z(t), \ldots, z^{(n)}(t)\right)\right) \tag{4.3a}$$

$$\triangleq \psi_{\mathrm{y}}\left(z(t), \ldots, z^{(q)}(t)\right), \quad \text{with } q = n - r. \tag{4.3b}$$

Equation (4.3b) is the output map in dependence upon z. The integer $r = n - q$ is the relative degree which determines the order of the IMC filter F.

As an intermediate step, assume that requirements on the behaviour of the closed-loop system are given in terms of a desired trajectory z_{d} for the flat output z. Further, assume that z_{d} is $(n-1)$-times continuously differentiable (i. e., $z \in C^{n-1}$) and that all derivatives ${z_{\mathrm{d}}}^{(i)}$ with $i = 0, \ldots, n$ are known. Then, the control input $u(t)$ can be determined by Eq. (4.2b) as

$$u(t) = \psi_{\mathrm{u}}\left(z_{\mathrm{d}}(t), \ldots, {z_{\mathrm{d}}}^{(n)}(t)\right), \tag{4.4}$$

which is an algebraic equation whose arguments are the known functions $z_{\mathrm{d}}, \dot{z}_{\mathrm{d}}, \ldots, {z_{\mathrm{d}}}^{(n)}$. For the purpose of this thesis, the main consequence of the flatness property is a perfect feedforward controller with respect to the flat output z:

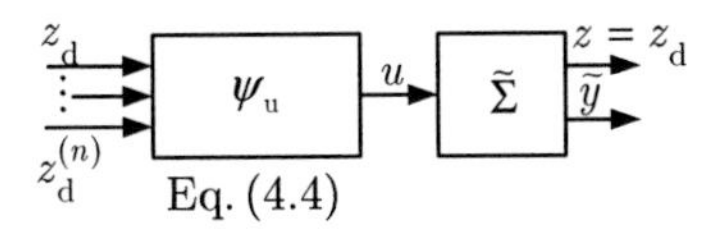

Fig. 4.1: Implementation of the flatness-based feedforward control structure.

Corollary 4.1 (Perfect Feedforward Controller for the Flat Output). *The system $\tilde{\Sigma}$ follows a given n-times differentiable trajectory $z_d(t) \in C^{n-1}$ with known derivatives ${z_d}^{(i)}(t)$ $(i = 1, \ldots, n)$* exactly

$$z(t) = z_d(t), \tag{4.5}$$

if the control input $u(t)$ given by Eq. (4.4) is used and

$$ {z_d}^{(i)}(0) = L_f^i \Phi(\boldsymbol{x}(0)), \quad i = 0, \ldots, n-1 \tag{4.6}$$

holds.

Equation (4.6) ensures that the desired trajectory z_d matches the initial state $\boldsymbol{x}(0)$ of the model $\tilde{\Sigma}$. Figure 4.1 shows the implementation of the resulting flatness-based feedforward control structure.

However, since the model output $\tilde{y}$ and its flat output z are not necessarily identical, the following introduces a mapping between them.

Right inverse of flat systems. In the following, the goal is to map the signal $\tilde{y}_d$ (which is supplied by the IMC filter F) into the corresponding signal z_d of the flat model output. Then, the mapping from $\tilde{y}_d$ into z (denoted by the operator $F_{y \to z}$), composed with ψ_u, yields the right inverse $\tilde{\Sigma}^r$ which is part of the IMC controller.

The relationship between $\tilde{y}_d(t)$ and $z_d(t)$ is obtained by using Eq. (4.3)

$$\psi_y\left(z_d(t), \ldots, {z_d}^{(q)}(t)\right) = \tilde{y}_d(t), \quad \text{with } q = n - r \tag{4.7a}$$

together with the initial condition given in Eq. (4.6). Equation (4.7a) is a differential equation which needs to be solved for $z_d(t)$. The solution is stable and can be obtained numerically because the model is assumed to be minimum-phase[1].

[1] If $F_{y \to z}$ is an unstable operator then the right inverse $\tilde{\Sigma}^r$ would be unstable which would imply an NMP model $\tilde{\Sigma}$ (cf. Definition 3.6).

In the sense of feedforward control, the control law (4.4) requires the knowledge of all derivatives up to the n-th order, whereas a numerical solution of Eq. (4.7a) only yields ${z_\mathrm{d}}^{(q)}(t)$ as highest derivative. To this end, additional r derivatives of Eq. (4.7a) have to be determined by

$$\frac{d}{dt}\left(\psi_\mathrm{y}\left(z_\mathrm{d}(t),\ldots,{z_\mathrm{d}}^{(q)}(t)\right)\right)=\dot{\tilde{y}}_\mathrm{d}(t) \tag{4.7b}$$

$$\vdots$$

$$\frac{d^r}{dt^r}\left(\psi_\mathrm{y}\left(z_\mathrm{d}(t),\ldots,{z_\mathrm{d}}^{(q)}(t)\right)\right)=\tilde{y}_\mathrm{d}^{(r)}(t). \tag{4.7c}$$

A numerical solution of the differential equation (4.7a) yields the required differentiations of the flat output from z_d to ${z_\mathrm{d}}^{(q)}$. The remaining higher derivatives can be obtained by the algebraic relationships (4.7b) to (4.7c) (and all in between) by iteratively inserting the solutions z_d to ${z_\mathrm{d}}^{(q)}$ into the equations from first to last and each time solving for the highest derivative of the flat output z_d. The equations (4.7b) to (4.7c) are algebraic as each only has one unknown variable (namely its highest derivative for z_d) as all lower derivatives of z_d are known from previous iterations or the differential equation (4.7a).

Remark 4.1 (Order of a flatness-based IMC)**.** From the above one finds that the order of the right inverse (i. e., the number of necessary integrators) is $q = n - r$ which results in an order of a flatness-based IMC of n.

For a given signal $\tilde{y}_\mathrm{d}^{(r)}$, solving the differential and algebraic equations (4.7) for z_d is denoted by the mapping

$$F_{\mathrm{y}\to\mathrm{z}}:\widetilde{\mathcal{Y}}\to\mathcal{Z}, \tag{4.8}$$

whose implementation must also deliver all derivatives of z_d up to n-th order.

The right inverse $\widetilde{\Sigma}^\mathrm{r}$ of a flat system is given by the composition

$$\widetilde{\Sigma}^\mathrm{r}=\psi_\mathrm{u}\circ F_{\mathrm{y}\to\mathrm{z}} \tag{4.9}$$

which is described by Eqns. (4.7c) and (4.4).

IMC controller. The following theorem proposes the flatness-based internal model controller.

Theorem 4.1 (Flatness-based IMC). *If the plant Σ can be represented by a stable minimum-phase flat model $\widetilde{\Sigma}$, a nonlinear IMC controller Q is described by*

$$k_r \tilde{y}_{\mathrm{d}}^{(r)} + k_{r-1} \tilde{y}_{\mathrm{d}}^{(r-1)} + \ldots + \tilde{y}_{\mathrm{d}} = \tilde{w} \tag{4.10a}$$

$$\begin{aligned} \psi_{\mathrm{y}}\left(z_{\mathrm{d}}(t), \ldots, z_{\mathrm{d}}{}^{(q)}(t)\right) &= \tilde{y}_{\mathrm{d}}(t), \quad \textit{with } q = n - r \\ &\vdots \\ \frac{d^r}{dt^r}\left(\psi_{\mathrm{y}}\left(z_{\mathrm{d}}(t), \ldots, z_{\mathrm{d}}{}^{(q)}(t)\right)\right) &= \tilde{y}_{\mathrm{d}}^{(r)}(t) \end{aligned} \tag{4.10b}$$

$$u = \psi_{\mathrm{u}}\left(z_{\mathrm{d}}, \ldots, z_{\mathrm{d}}{}^{(n)}\right), \tag{4.10c}$$

with the filter coefficients k_i $(i = 1, \ldots, r)$ and the function ψ_{u} resulting due to the flatness property (4.2b) *of the model. The Eqns.* (4.10b) *represent the operator $F_{\mathrm{y}\to\mathrm{z}}$, defined in Eq.* (4.8). *The initial conditions for the differential equations* (4.10a) *and* (4.10b) *are given in Eqns.* (3.13) *or* (4.6), *respectively.*

Equations (4.10a) and (4.10b) represent the dynamics of Q and the algebraic relationship (4.10c) maps the states of the dynamics into the plant input u. The differential equation (4.10a) follows directly from the transfer functions (3.34) and (3.35) of the IMC filter F. The parameters k_i with $i = 1, \ldots, r$ represent the degree-of-freedom of Q. The differential equation (4.10b) together with the output map (4.10c) is the right inverse $\widetilde{\Sigma}^{\mathrm{r}}$ of the flat model $\widetilde{\Sigma}$.

The resulting IMC control loop shown in Fig. 4.2 is robustly stable (cf. Theorem 3.2) and if stability is achieved, the closed-loop has zero steady-state offset (cf. Property 3.3). The resulting closed-loop behaviour depends on the choice of the k_i and the modelling errors. According to Step 3 of Algorithm 3.2, the IMC control loop can be implemented without the explicit model as shown in Fig. 4.2. The important improvement obtained by the method presented in this thesis is the fact that this application of the IMC principle is not based on feedback linearisation. The controller described by Theorem 4.1 achieves robust tracking performance and stability directly for the nonlinear plant.

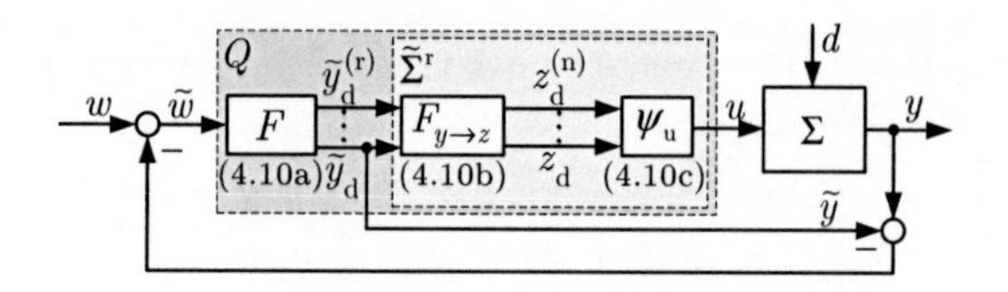

Fig. 4.2: Flatness-based IMC control structure.

Example 4.1 (Flatness-based IMC):
Consider the plant model

$$\tilde{\Sigma}: \begin{aligned} \dot{x}_1 &= -x_1^3 + x_2 \\ \dot{x}_2 &= -x_2^3 - x_1 x_2 + \sigma u \end{aligned} \qquad \boldsymbol{x}(0) = \begin{bmatrix} 2, & 0 \end{bmatrix}^T \tag{4.11a}$$

$$\tilde{y} = x_1 + x_2, \tag{4.11b}$$

where $\sigma \neq 0$ is an uncertain parameter with the nominal value $\sigma = 1$.

The first time-derivative of the output $\tilde{y}$ depends explicitly on the input u. Thus, the system (4.11), has a relative degree of $r = 1$. The variable $z = x_1$ is a flat output, since

$$z = x_1 \qquad \Leftrightarrow \qquad x_1 = z = \psi_{11}(z) \tag{4.12a}$$

$$\dot{z} = \dot{x}_1 = -x_1^3 + x_2 \qquad \Leftrightarrow \qquad x_2 = \dot{z} + z^3 = \psi_{12}(z, \dot{z}) \tag{4.12b}$$

$$\ddot{z} = -3x_1^2\dot{x}_1 + \dot{x}_2 = -3x_1^2(-x_1^3 + x_2) - x_2^3 - x_1 x_2 + \sigma u$$

$$\Leftrightarrow \sigma u = \ddot{z} + (\dot{z} + z^3)^3 + \dot{z}(z + 3z^2) + z^4 \tag{4.12c}$$

holds. Consequently, for $\sigma = 1$, the flat feedforward control law (see Eq. (4.4)) can be determined as:

$$u = \psi_u(z_d, \dot{z}_d, \ddot{z}_d) = \ddot{z}_d + (\dot{z}_d + z_d^3)^3 + \dot{z}_d(z_d + 3z_d^2) + z_d^4 \tag{4.13}$$

The mapping $F_{y \to z}$ is constructed according to the procedure described in Section 4.1:

$$z_d + z_d^3 + \dot{z}_d = \tilde{y}_d \tag{4.14a}$$

$$\dot{z}_d + 3z_d^2\dot{z}_d + \ddot{z}_d = \dot{\tilde{y}}_d. \tag{4.14b}$$

Equation (4.14a) is a nonlinear differential equation for z with input $\tilde{y}_d$. It is stable and, thus, can be solved numerically with the initial conditions $z_d(0) = x_1(0)$ and $\dot{z}_d(0) = \dot{x}_1(0)$. To obtain the additional derivative $\ddot{z}_d$, Eq. (4.14b) is used. The structure of the required dynamic model $F_{y \to z}$ is presented in Fig. 4.3. The IMC filter F has to be implemented as a first-

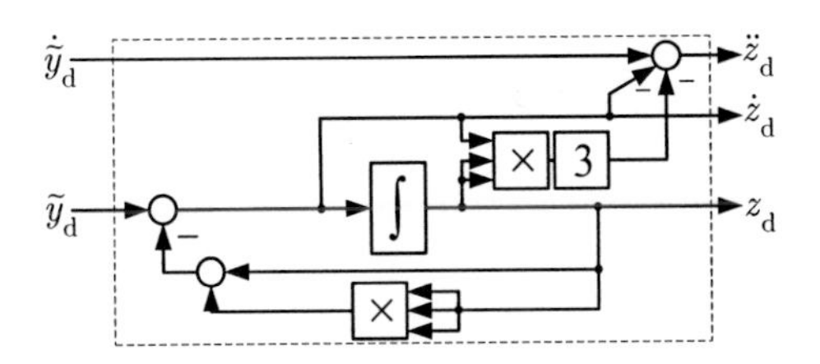

Fig. 4.3: $F_{y\to z}$ from Eq. (4.14).

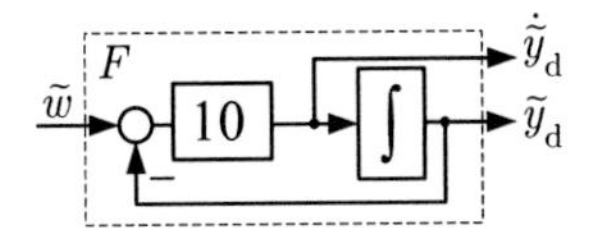

Fig. 4.4: First-order IMC filter F.

order filter (Fig. 4.4). The filter pole is chosen as $\lambda = -10$ and the initial condition set to $\tilde{y}_{\mathrm{d}}(0) = x_1(0) + x_2(0) = 2$.

Finally, the flat IMC controller Q can be implemented according to the structure in Fig. 4.2, namely as the interconnection of the first-order filter F (see Fig. 4.4), the mapping $F_{y\to z}$ shown in Fig. 4.3, and the feedforward control law (4.13).

Figure 4.5 shows the performance of the closed-loop. The reference w is initially zero and modified by two steps occurring at $t = 1\mathrm{s}$ and $t = 2\mathrm{s}$. It can be seen that for the nominal case $\sigma = 1$, exact tracking of the generated trajectory $\tilde{y}_{\mathrm{d}}$ is achieved, and that for a changed parameter σ robust control performance is provided. ■

4.2 IMC of SISO Systems in Input/Output Normal Form

In this section, the I/O normal form of an input-affine system is exploited to obtain a right inverse (Step 1 of Algorithm 3.2).

I/O normal form. A scalar input-affine model $\tilde{\Sigma}$ is defined by

$$\tilde{\Sigma}: \quad \begin{aligned} \dot{\boldsymbol{x}} &= \boldsymbol{f}(\boldsymbol{x}) + \boldsymbol{g}(\boldsymbol{x})\,u, \quad \boldsymbol{x}(0) = \boldsymbol{x}_0 \\ \tilde{y} &= h(\boldsymbol{x}). \end{aligned} \tag{4.15}$$

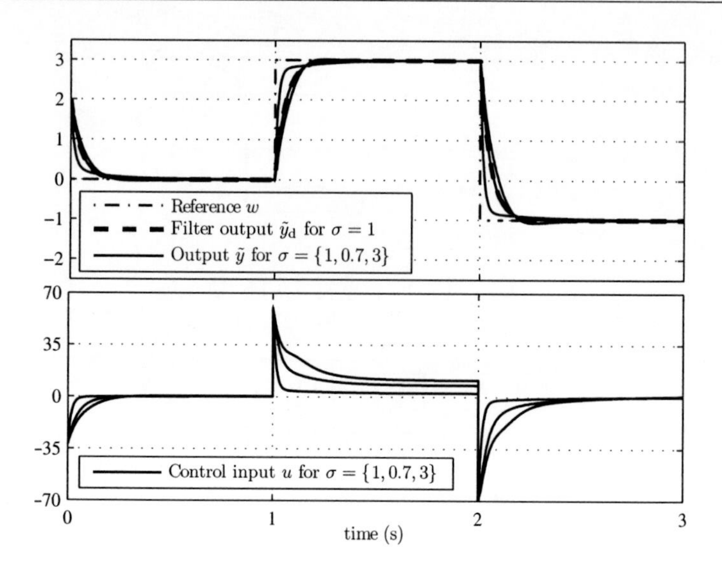

Fig. 4.5: Step responses of the closed-loop system of Example 4.1.

It is assumed that the elements of the vector fields $\boldsymbol{f}$ and $\boldsymbol{g}$ are analytic functions of their arguments. The following reviews the I/O normal form of nonlinear systems.

It is assumed that the model $\widetilde{\Sigma}$ can be transformed with $\boldsymbol{x}^*$ as the new coordinates defined by (cf. [51])

$$
\begin{aligned}
x_1^* &= \tilde{y} = h(\boldsymbol{x}) && = \phi_1(\boldsymbol{x}) && \text{(4.16a)}\\
x_2^* &= \dot{\tilde{y}} = \dot{x}_1^* && = L_f h(\boldsymbol{x}) = \phi_2(\boldsymbol{x}) && \text{(4.16b)}\\
&\vdots && \vdots \\
x_r^* &= \tilde{y}^{(r-1)} = \dot{x}_{r-1}^* && = L_f^{r-1} h(\boldsymbol{x}) = \phi_r(\boldsymbol{x}) && \text{(4.16c)}\\
x_{r+1}^* &= \phi_{r+1}(\boldsymbol{x}) && && \text{(4.16d)}\\
&\vdots \\
x_n^* &= \phi_n(\boldsymbol{x}). && && \text{(4.16e)}
\end{aligned}
$$

The functions $\phi_i(\boldsymbol{x})$ for $i = r+1, \ldots, n$ can be chosen arbitrarily with the requirement that the map $\boldsymbol{\phi}(\boldsymbol{x}) = [\phi_1, \cdots, \phi_n]^T$ is a diffeomorphism[2].

[2] A diffeomorphism is defined as a differentiable map between manifolds which has

Locally, a diffeomorphism is required to fulfil

$$\text{Rank}\left(\frac{\partial \boldsymbol{\phi}(\boldsymbol{x})}{\partial \boldsymbol{x}}\right) = n. \tag{4.17}$$

After r differentiations of $\tilde{y}$, the input u appears for the first time explicitly

$$\tilde{y}^{(r)} = L_f^r h(\boldsymbol{x}) + \underbrace{L_g L_f^{r-1} h(\boldsymbol{x})}_{\neq 0}\, u. \tag{4.18}$$

The number r is the relative degree of $\widetilde{\Sigma}$ with respect to its output $\tilde{y}$. It is assumed that $r < n$ holds. For $r = n$ the output $\tilde{y}$ is a flat output and the result of the preceding section solves the problem. For $r = 0$ the output $\tilde{y}$ explicitly depends on u, resulting in a direct feedthrough ($\tilde{y} = h(\boldsymbol{x}, u)$) for which the following procedure is still applicable.

In the new coordinates $\boldsymbol{x}^*$, $\widetilde{\Sigma}$ can be written in the I/O normal form

$$\begin{bmatrix} \dot{x}_1^* \\ \dot{x}_2^* \\ \vdots \\ \dot{x}_{r-1}^* \end{bmatrix} = \begin{bmatrix} 0 & 1 & 0 & \cdots & 0 \\ 0 & 0 & 1 & \ddots & 0 \\ \vdots & & & \ddots & 0 \\ 0 & \cdots & & & 1 \end{bmatrix}_{(r-1 \times r)} \cdot \begin{bmatrix} x_1^* \\ x_2^* \\ \vdots \\ x_r^* \end{bmatrix} \tag{4.19a}$$

$$\dot{x}_r^* = a(\boldsymbol{x}^*) + b(\boldsymbol{x}^*)u, \quad b(\boldsymbol{x}^*) \neq 0 \tag{4.19b}$$

$$\begin{bmatrix} \dot{x}_{r+1}^* & \cdots & \dot{x}_n^* \end{bmatrix}^T = \boldsymbol{p}(\boldsymbol{x}^*) + \boldsymbol{q}(\boldsymbol{x}^*)u \tag{4.19c}$$

with the initial condition $\boldsymbol{x}^*(0) = \boldsymbol{\phi}(\boldsymbol{x}(0))$ where $a(\boldsymbol{x}^*), b(\boldsymbol{x}^*), \boldsymbol{p}(\boldsymbol{x}^*)$ and $\boldsymbol{q}(\boldsymbol{x}^*)$ are given in new coordinates $\boldsymbol{x}^*$ by (see e.g., [51, 60, 61])

$$\begin{aligned} a(\boldsymbol{x}^*) &= L_f^r h \circ \boldsymbol{\phi}^{-1}(\boldsymbol{x}^*) \\ b(\boldsymbol{x}^*) &= L_g L_f^{r-1} h \circ \boldsymbol{\phi}^{-1}(\boldsymbol{x}^*) \\ p_i(\boldsymbol{x}^*) &= L_f \phi_i \circ \boldsymbol{\phi}^{-1}(\boldsymbol{x}^*) \\ q_i(\boldsymbol{x}^*) &= L_g \phi_i \circ \boldsymbol{\phi}^{-1}(\boldsymbol{x}^*) \end{aligned}$$

for $i = r+1, \cdots, n$.

Right inverse of I/O-linearisable systems. From the I/O normal form, a right inverse can be obtained: Let the input be defined by

$$u(t) = \frac{1}{b(\boldsymbol{x}^*(t))}\left(-a(\boldsymbol{x}^*(t)) + \nu(t)\right), \tag{4.20}$$

a differentiable inverse.

where $\nu(t)$ is a new input, that changes Eq. (4.19b) into $\dot{x}_r^*(t) = \nu(t)$ and from Eq. (4.16) it follows that

$$\dot{x}_r^*(t) = \nu(t) = \tilde{y}^{(r)}(t) \tag{4.21}$$

holds. Demanding $\tilde{y} \overset{!}{=} \tilde{y}_\text{d}$, Eqns. (4.19b) and (4.19c) can be rewritten as

$$u = \frac{1}{b(\boldsymbol{x}^*)}\left(-a(\boldsymbol{x}^*) + \tilde{y}_\text{d}^{(r)}\right) \tag{4.22}$$

$$\left[\dot{x}_{r+1}^*, \quad \cdots, \quad \dot{x}_n^*\right]^T = \overline{\boldsymbol{p}}(\boldsymbol{x}^*) + \overline{\boldsymbol{q}}(\boldsymbol{x}^*)\tilde{y}_\text{d}^{(r)}. \tag{4.23}$$

Additionally, it is known from Eq. (4.16) that a part of the transformed state vector $\boldsymbol{x}^*$ is given by the IMC filter F

$$[x_1^*, \ldots, x_r^*]^T = [\tilde{y}_\text{d}, \ldots, \tilde{y}_\text{d}^{(r-1)}]^T. \tag{4.24}$$

In order to use the control law (4.22), the value of the unknown states $[x_{r+1}^*, \ldots, x_n^*]^T$ needs to be determined. It is proposed to do this by numerically solving the differential equation (4.23), which is possible due to the restriction to minimum-phase models. The dynamics posed from Eq. (4.23) is called the internal dynamics.

It seems important to clarify the relationship between the internal dynamics and the zero dynamics. Although the notion of zero dynamics is not directly used for controller design, it is discussed briefly in the following remark for the sake of completeness.

Remark 4.2 (Zero dynamics [51]). The notion of zero dynamics means the case of the trajectory of the internal dynamics that is generated by the input function $u_0 = \gamma(\boldsymbol{x}^*)$ which keeps the output $\tilde{y}$ at zero for all time $t > 0$ (assuming matching initial conditions), which can be expressed as $x_1^*(t), \ldots, x_r^*(t) = 0$, $\forall t \geq 0$.

The zero dynamics are then defined as the dynamics of the states of the internal dynamics (4.23) $x_{r+1}^*, \ldots, x_n^*$ under the influence of u_0 and $x_1^*, \ldots, x_r^* = 0$. Despite their interesting system theoretical properties, the zero dynamics are not exploited for an IMC control design. Therefore, the notion of the zero dynamics does not play a direct role in IMC design.

In order for the zero dynamics to be important in control design, the output $\tilde{y}$ has be defined as the difference of the output of a reference system and the output of the plant. Then, it is desired to maintain this output $\tilde{y}$ constantly at zero. Such a situation is discussed, for example, in [65].

> With the above, the right inverse $\widetilde{\Sigma}^{\mathrm{r}}$ for systems which can be transformed into the I/O normal form is given by Eqns. (4.22)-(4.24).

The implementation of the right inverse $\widetilde{\Sigma}^{\mathrm{r}}$ of I/O linearisable systems is displayed in Fig. 4.6. Note that the input vector signal $[\tilde{y}_{\mathrm{d}}, \ldots, \tilde{y}_{\mathrm{d}}^{(r)}]^T$ is delivered by the IMC filter F.

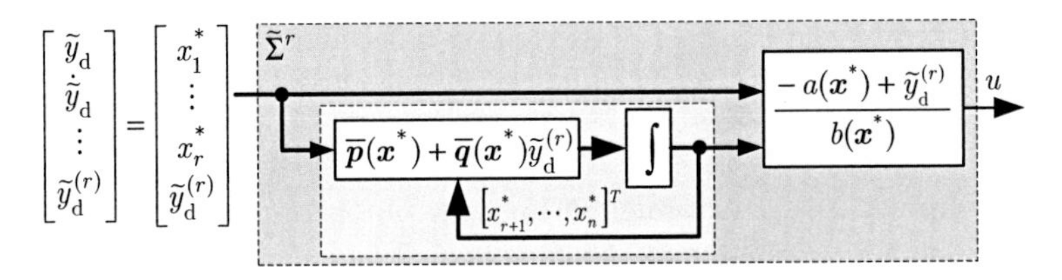

Fig. 4.6: Right inverse $\widetilde{\Sigma}^{\mathrm{r}}$ of a system in I/O normal-form (4.19). Bold arrows indicate vector signals.

IMC controller. The following theorem summarises the results of this section.

> **Theorem 4.2 (IMC for Input-Affine Systems).** *If the plant* Σ *can be represented by a stable minimum-phase input-affine model* $\widetilde{\Sigma}$, *which can be transformed into the I/O normal form* (4.19), *a nonlinear IMC controller* Q *is given by*
>
> $$\begin{aligned} k_r \tilde{y}_{\mathrm{d}}^{(r)} + k_{r-1} \tilde{y}_{\mathrm{d}}^{(r-1)} + \ldots + \tilde{y}_{\mathrm{d}} &= \tilde{w} \\ [x_1^*, \ldots, x_r^*]^T &= [\tilde{y}_{\mathrm{d}}, \ldots, \tilde{y}_{\mathrm{d}}^{(r-1)}]^T \end{aligned} \tag{4.25a}$$
>
> $$\left[\dot{x}_{r+1}^*, \quad \cdots, \quad \dot{x}_n^*\right]^T = \overline{p}(x^*) + \overline{q}(x^*)\tilde{y}_{\mathrm{d}}^{(r)} \tag{4.25b}$$
>
> $$u = \frac{1}{b(x^*)}\left(-a(x^*) + \tilde{y}_{\mathrm{d}}^{(r)}\right), \tag{4.25c}$$
>
> *with the filter coefficients* k_i $(i = 1, \ldots, r)$ *and the relationships* (4.25b) *and* (4.25c) *which result from the I/O normal form* (4.19) *of the model and the introduction of the new input in* (4.20). *The initial conditions for the differential equations* (4.25a) *and* (4.25b) *are given by* $x^*(0) = \Phi(x(0))$.

Eqns. (4.25a) and (4.25b) represent the dynamics of Q and the algebraic relationship (4.25c) maps the states of the dynamics into the plant input u. The parameters k_i with $i = 1, \dots, r$ represent the degree-of-freedom of Q and can be chosen freely with the limitation that the differential equation (4.25a) must be stable. The differential equation (4.25a) follows directly from the transfer function (3.34) of the IMC filter F.

The resulting IMC control loop (Fig. 3.8) relies on output feedback, is robustly stable (Theorem 3.2), and if stability is achieved, the closed-loop has zero steady-state offset (Property 3.3). The resulting closed-loop behaviour depends on the choice of the k_i and the modelling error.

Example 4.2 (IMC of an I/O linearisable system):
Consider the input-affine model

$$\begin{aligned} \dot{x}_1 &= -\sigma x_1^3 + x_2 + u \\ \dot{x}_2 &= -x_2^3 - x_1 x_2 + u \end{aligned} \qquad \boldsymbol{x}(0) = \begin{bmatrix} -3.5, & 3 \end{bmatrix}^T \tag{4.26a}$$

$$\tilde{y} = x_1 + x_2, \tag{4.26b}$$

where σ is an uncertain parameter and the IMC controller is designed for the nominal case $\sigma = 1$.

In the I/O normal form, the first transformed state x_1^* is equal to the output:

$$x_1^* = \tilde{y} = x_1 + x_2 = \Phi_1(\boldsymbol{x}). \tag{4.27}$$

Its first derivative

$$\dot{x}_1^* = \dot{\tilde{y}} = -x_1^3 + x_2 - x_2^3 - x_1 x_2 + 2u \tag{4.28}$$

depends on u, thus the relative degree is $r = 1$. Employing Eq. (4.22) one finds

$$u = \frac{1}{2}\left(x_1^3 - x_2 + x_2^3 + x_1 x_2 + \nu\right), \tag{4.29}$$

where ν is the new input. Equation (4.29) is obtained by setting ν to the differentiated IMC filter output: $\nu = \dot{\tilde{y}}_\mathrm{d}$. Choose

$$x_2^* = x_1 - x_2 = \Phi_2(\boldsymbol{x}) \tag{4.30}$$

as state for the internal dynamics, which consequently is determined as

$$\dot{x}_2^* = -x_1^3 + x_2 + x_2^3 + x_1 x_2. \tag{4.31}$$

Since the relative degree is $r = 1$, the first-order filter shown in Fig. 4.4 can be used. Equations (4.29) and (4.31) are still expressed in the original

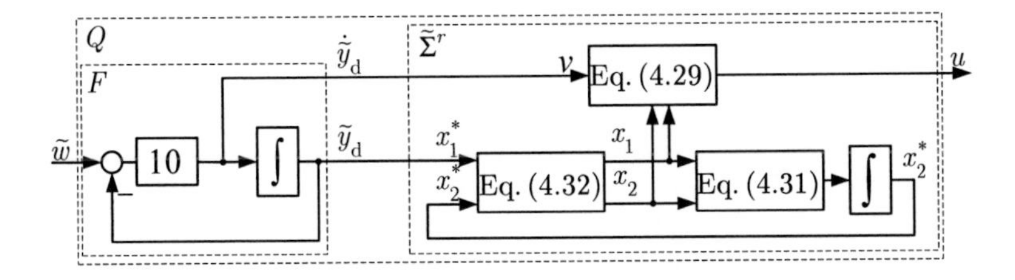

Fig. 4.7: Structure of the controller Q.

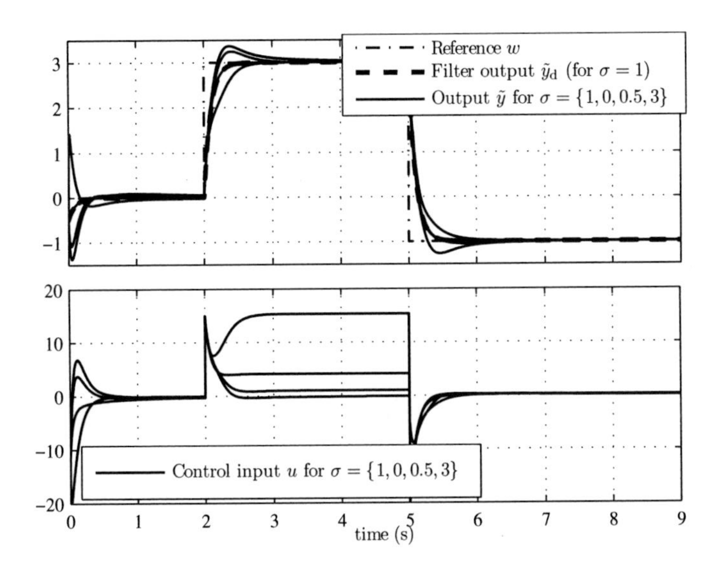

Fig. 4.8: Simulation results of Example 4.2.

coordinates. For implementation, these coordinates can be substituted by the inverse coordinate transformation, given by

$$x_1 = \frac{1}{2}(x_1^* + x_2^*) \qquad\qquad x_2 = \frac{1}{2}(x_1^* - x_2^*). \tag{4.32}$$

The structure of the complete resulting IMC controller Q is presented in Fig. 4.7. Figure 4.8 displays the simulation results for the system (4.26a), (4.26b) controlled by the IMC controller shown in Fig. 4.7. It gives the response of the closed-loop system subject to two consecutive steps in w at time $t = 2$s and $t = 5$s. The system output y is plotted for the nominal case ($\sigma = 1$) as well as for $\sigma = \{0, 0.5, 3\}$. It is shown that despite this significant modelling error, the closed-loop behaviour robustly yields zero steady-state offset. ■

Example 4.3 (Comparison to Exact I/O Linearisation):
The example above is continued to portray the differences between an IMC based on the I/O normal form and an exact I/O linearisation. An I/O linearisation for the system (4.26) is performed for the nominal case $\sigma = 1$ according to [51]: The input law (4.29) remains the same with the exception that the states x_1 and x_2 have to be measured from the plant (4.26). However, the new input ν is now used to establish the *linear* error dynamics

$$\nu = \dot{\tilde{y}}_{\mathrm{d}} - c_1(y - \tilde{y}_{\mathrm{d}}). \tag{4.33}$$

The pole of the error dynamics is set to the pole of the IMC filter:

$$c_1 = -\lambda. \tag{4.34}$$

An exact I/O linearisation only stabilises the plant around an *a priori given* and r-times differentiable trajectory $\tilde{y}_{\mathrm{d}}$. In this example, this trajectory is generated by sending the reference signal w through the filter F and its outputs $\dot{\tilde{y}}_{\mathrm{d}}$ and $\tilde{y}_{\mathrm{d}}$ are then assumed to be the given trajectory. In conclusion, the exact I/O linearisation requires state feedback and consists of Eqns. (4.29), (4.33) and (4.34).

For the nominal case ($\sigma = 1$), simulation results are identical to the proposed nonlinear IMC controller presented in Example 4.2. However, simulations with varying parameter σ (assuming a given reference signal w and initial condition $\boldsymbol{x}_0$) show that the exact I/O linearisation becomes unstable for $\sigma \leq 0.73$, while the IMC loop remains stable for all positive values of σ and even for negative values $\sigma \geq -0.4$.

Figure 4.9 shows a comparison for $\sigma = 0.72$ between the nonlinear IMC controller obtained in Example 4.2 and the exact I/O linearisation as developed here. The result shows that, here, the exact I/O linearisation is not as robust as IMC. The I/O linearised (but perturbed) plant shows unstable behaviour between $4s \leq t \leq 5$s while the output of the IMC controlled plant shows an almost nominal behaviour.

In this example, the nonlinear IMC is to be preferred over the exact I/O linearisation since it only relies on output feedback instead of state feedback, possesses zero steady-state offset in the presence of modelling errors and shows larger robustness boundaries. Simulation studies with several other examples also indicate that IMC seems generally more robust than exact linearisation methods. ■

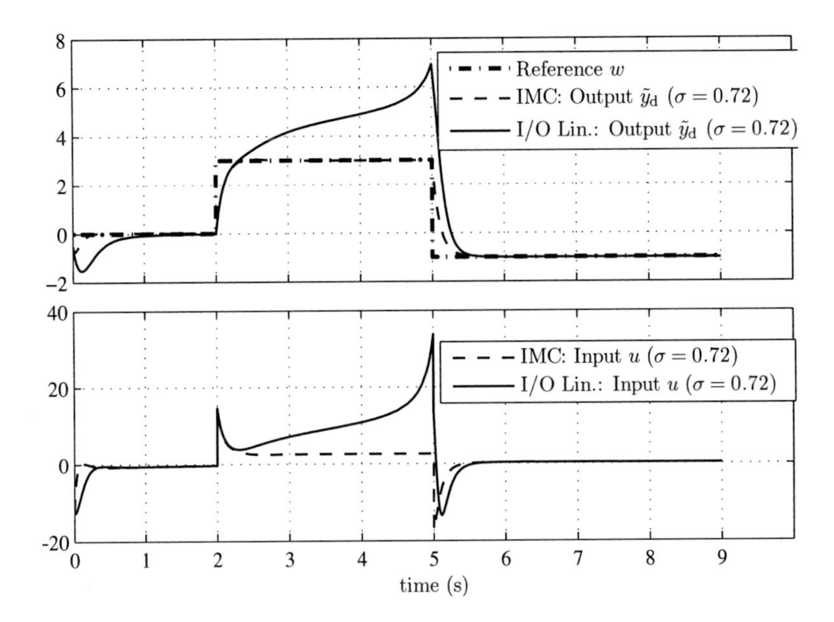

Fig. 4.9: Simulation results of Example 4.3.

4.3 Summary

The main contributions of this chapter are Theorem 4.1 and Theorem 4.2. The theorems give the necessary equations for designing an IMC controller Q for flat or input-affine systems, respectively. Essentially, right inverses for these system classes were introduced and combined with the IMC filter as introduced in Section 3.5.1.

Therefore, this chapter can be interpreted as the link between the rather abstract operator-oriented view on nonlinear IMC, as presented in Chapter 3, and the nonlinear IMC design for systems given in their state-space form.

5. EXTENSIONS OF THE BASIC PRINCIPLE

The main contribution of this chapter is the extension of the system class. In the preceding chapters, the plant model was restrained in such that a well-defined relative degree was necessary and arbitrarily large input signals were allowed. Moreover, the model was assumed to be MP (i. e., a stable inverse was required).

All of these assumptions are dropped in this chapter. In Section 5.1 input constraints are introduced to accommodate the fact that virtually all real plants only accept a limited range of the inputs u. In Section 5.2, the assumption of a well-defined relative degree r is dropped and replaced by the much weaker assumption of model invertibility (cf. Definition 3.5). As a result, the model may be singular at some points; that is, it may drop or lose its relative degree completely. As the example of the two-stage turbocharged engine in Chapter 7 shows, real plants may exhibit such a behaviour.

The minimum-phase assumption is removed in Section 5.3. A method is proposed with which a perfect (and therewith unstable) inverse can be employed by using an appropriate IMC filter F. It is shown how such an IMC controller can be built to be internally stable.

Finally, Sections 5.4 and 5.5 introduce some necessary amendments to the design method which are necessary for the applications in Part II. These amendments are the introduction of measured disturbances and treating simple quadratic MIMO systems.

5.1 *Input Constraints*

In all existing plants (e. g., machinery, chemical processes), the accessible inputs $u \in \mathcal{U}$ are limited. This is addressed by the input-function space

$$\mathcal{U} = \{u \mid u_{\min} \leq u(t) \leq u_{\max},\ \forall t\}\,. \tag{5.1}$$

The control problem to be solved involves respecting these input constraints. This means that the output u of the right inverse $\widetilde{\Sigma}^{\mathrm{r}}$ must fulfil

Eq. (5.1).

Main idea. From the nonlinear IMC design of Chapter 3 one finds the following: Definition 3.4 shows that the right inverse $\widetilde{\Sigma}^{\mathrm{r}}$ is only defined for input signals $\tilde{y}_{\mathrm{d}}$ in the output-function space (i. e., $\tilde{y}_{\mathrm{d}} \in \widetilde{\mathcal{Y}}$) of the model $\widetilde{\Sigma}$. The IMC filter F is responsible for mapping arbitrary input signals $\tilde{w} \in \mathcal{L}_\infty$ into signals that the right inverse "understands" (i. e., $\tilde{y}_{\mathrm{d}} \in \widetilde{\mathcal{Y}}$).

Therefore, input constraints can be considered by assuring that the IMC filter F only produces signals $\tilde{y}_{\mathrm{d}}$ which lie in the range of the model $\widetilde{\Sigma}$ (i. e., $\tilde{y}_{\mathrm{d}} \in \widetilde{\mathcal{Y}}$) reached from the domain of the limited inputs of Eq. (5.1). Thus, it is proposed to change F to produce only such trajectories $\tilde{y}_{\mathrm{d}} \in \widetilde{\mathcal{Y}}$ that can be obtained with permissible inputs $u \in \mathcal{U}$.

Applying Proposition 3.2 shows that the output-function space $\widetilde{\mathcal{Y}}$ is influenced by the input-function space $\mathcal{U}$ via the relative degree r. Hence, a limitation on the plant input u directly relates to a limitation on the highest derivative $\tilde{y}^{(r)}$ of the model output $\tilde{y}$. With the substitution $\tilde{y} \overset{!}{=} \tilde{y}_{\mathrm{d}}$ (cf. (3.30)) one finds the solution to the problem.

Theorem 5.1 (Input Limitation with IMC). *Consider a model $\widetilde{\Sigma}$ with a relative degree r. If the highest derivative $\tilde{y}_{\mathrm{d}}^{(r)}$ of the filter output $\tilde{y}_{\mathrm{d}}$ is limited to*

$$\tilde{y}_{\mathrm{d,min}}^{(r)}(t) \leq \tilde{y}_{\mathrm{d}}^{(r)}(t) \leq \tilde{y}_{\mathrm{d,max}}^{(r)}(t) \tag{5.2a}$$

with

$$\begin{aligned} \tilde{y}_{\mathrm{d,min}}^{(r)}(t) &= \min_{u \in \mathcal{U}} \varphi(\boldsymbol{x}(t), u) \\ \tilde{y}_{\mathrm{d,max}}^{(r)}(t) &= \max_{u \in \mathcal{U}} \varphi(\boldsymbol{x}(t), u) \end{aligned} \tag{5.2b}$$

and $\varphi(\boldsymbol{x}(t), u)$ as given in Definition 3.3 then the output u of the right inverse $\widetilde{\Sigma}^{\mathrm{r}}$ will always respect the limitation (5.1) *(i. e., it fulfils $u \in \mathcal{U}$) regardless of the setpoint w, disturbance d and modelling error.*

Proof. With a relative degree r and at a state $\boldsymbol{x}(t)$, Eq. (3.11) (given in Definition 3.3) the r-th output derivative $\tilde{y}^{(r)}(t)$ satisfies

$$\tilde{y}^{(r)}(t) = \varphi(\boldsymbol{x}(t), u(t)).$$

Since $\varphi(\boldsymbol{x}, u)$ is analytic in u (see the proof of Proposition 3.2), the contiguous range in u, given in Eq. (5.1), is mapped into a contiguous range

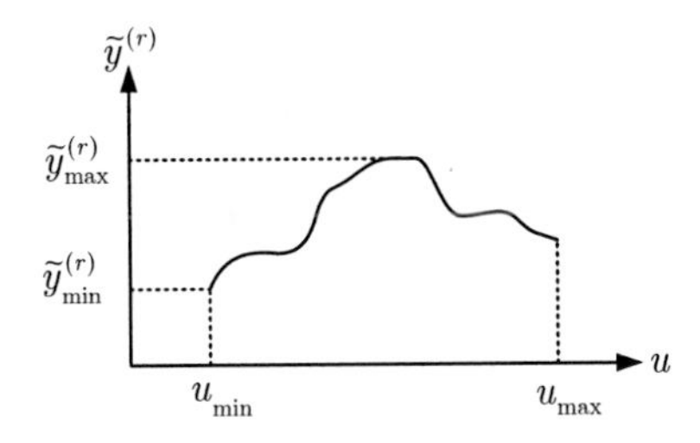

Fig. 5.1: Example of the mapping $\tilde{y}^{(r)}(t) = \varphi(\boldsymbol{x}(t), u(t))$ assuming a given $\boldsymbol{x}(t)$.

in $\tilde{y}^{(r)}$.

Figure 5.1 illustrates a mapping between u and $\tilde{y}^{(r)}$ at a given $\boldsymbol{x}(t)$. It shows the above properties of the function $\varphi(\cdot, u)$, namely that a contiguous range in u is mapped into a contiguous range in $\tilde{y}^{(r)}$, that $\tilde{y}^{(r)}$ over u is a smooth function (it is analytic), and that this mapping is not necessarily invertible.

It follows that for each $\tilde{y}^{(r)}(t)$ in the set $[\tilde{y}^{(r)}_{\text{d,min}}(t), \tilde{y}^{(r)}_{\text{d,min}}(t)]$, the algebraic relationship in Eq. (3.11) can be solved for a feasible input $u(t)$ in the permissible range $[u_{\text{min}}, u_{\text{max}}]$. From Eq. (3.30) it follows that the IMC filter F achieves $\tilde{y} = \tilde{y}_{\text{d}}$ and, thus, Eq. (3.11) is also valid for $\tilde{y} = \tilde{y}_{\text{d}}$. □

Implementation. It is proposed to implement the limitation (5.2a) by a saturation block in the SVF structure of the IMC filter F as shown in Fig. 5.2. This proposed implementation ensures that the filter F still

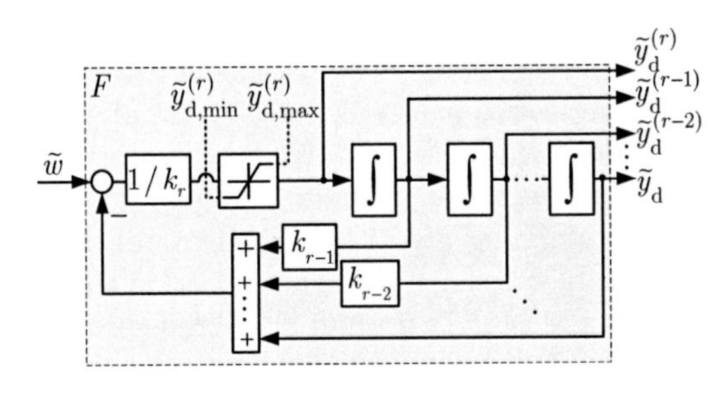

Fig. 5.2: Nonlinear limited IMC filter F to be used for input constraints.

achieves its steady-state value despite arbitrarily many intermitted limitations on $\tilde{y}_\mathrm{d}^{(r)}$, as long as no limitation occurs for $t \to \infty$.

Interpretation. The resulting *nonlinear* state-variable IMC filter F, shown in Fig. 5.2 can be interpreted as a control structure where the k_i $(i = 1, \ldots, r)$ can be interpreted as the gains of a state feedback controller for the plant which consists of the integrator chain. The saturation can then be interpreted as a modelling error of the integrator chain whose effect needs to be attenuated.

The function of the limited IMC filter F, shown in Fig. 5.2, is the following:

- If the input u is not in saturation then the behaviour of the saturated filter F (Fig. 5.2) is identical to the behaviour of the unsaturated filter given by Eq. (3.34). Hence, if not in saturation, the IMC filter yields the linear I/O behaviour as chosen by the designer.

- If in saturation, the IMC filter F produces that same output trajectory $\tilde{y}_\mathrm{d} = \tilde{y}$ that the model $\tilde{\Sigma}$ produces with the saturated input u. Then, the IMC filter F has a nonlinear I/O behaviour.

Literature. One concludes that this result and the design method of IMC yield a feedforward controller Q that can respect input constraints on-line. The design of nonlinear feedforward controllers is a topic of current research (see e. g., [40, 75, 94, 100]). To this end, several algorithms to design trajectories and their differentiations for use with a right inverse are proposed in the literature mentioned above. In [94], the authors propose the use of polynomials to generate $\tilde{y}_\mathrm{d}(t)$ and its necessary derivatives. In order to respect the input constraints, a numerical iteration is proposed where the slope of the polynomial is adjusted iteratively until the output u of the right inverse is within the acceptable input constraints. This approach does not meet the real-time demands in the automotive industry nor can it handle ramps in $w(t)$ nicely. In [40], the authors propose an offline trajectory planning algorithm which relies on the solution of a boundary value problem. This approach relies on the knowledge of a future setpoint from which the information of the second boundary can be obtained. It can handle input constraints well, but its application is restricted to problems where the setpoints are known a priori. Finally, [100] proposes a similar approach as the one discussed here with the difference that the trajectory $\tilde{y}_\mathrm{d}$ is not computed by a linear filter but rather is the output of a sliding mode controller. Although this idea is mathematically more involved than limiting an SVF, it does offer the possibility to respect additional constraints on, for example the model states.

In conclusion, the limited IMC filter F offers a computationally cheap and straightforward way to generate trajectories for nonlinear feedforward control that respect input constraints.

Finally, Fig. 5.3 shows the feedforward controller Q which can be used as an internal model controller. Note that not all states x_i in the state

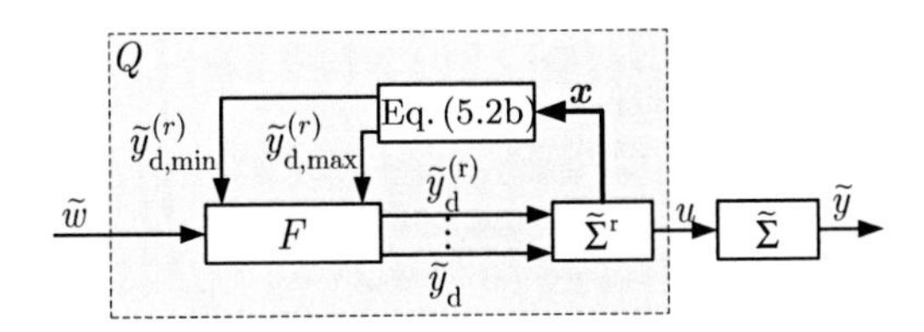

Fig. 5.3: Feedforward control structure using the input constraints as proposed here.

vector $\boldsymbol{x}$ as depicted in Fig. 5.3 need to be determined by the right inverse $\tilde{\Sigma}^{\mathrm{r}}$. For systems without internal dynamics, the limitations in Eq. (5.2b) can be computed using the information from the filter F only.

The actual calculation of $\boldsymbol{x}(t)$, which is used to determine the boundaries $\tilde{y}^{(r)}_{\mathrm{d,min}}(t)$ and $\tilde{y}^{(r)}_{\mathrm{d,max}}(t)$ using Eq. (5.2b), depends on the system class and is discussed in the following.

Input constraint with a flatness-based IMC. In case of a flatness-based IMC, the relationship Eq. (3.11) between the highest derivative $\tilde{y}^{(r)}_{\mathrm{d}}$ of the filter output $\tilde{y}_{\mathrm{d}}$ and the input u of the model can be obtained by using the flat output z.

Using Theorem 4.1 on page 73, Eq. (4.10c) can always[1] uniquely be solved for the highest derivative ${z_{\mathrm{d}}}^{(n)}$ of the flat output. Suppose the solution is written as a function ψ_{u}^{-1} of the lower derivatives and the input u

$$ {z_{\mathrm{d}}}^{(n)} = \psi_{\mathrm{u}}^{-1}(z_{\mathrm{d}}, \dot{z}_{\mathrm{d}}, \ldots, {z_{\mathrm{d}}}^{(n-1)}, u). \tag{5.3} $$

As an intermediate step, the permissible range of the highest derivative ${z_{\mathrm{d}}}^{(n)}$ is to be limited according to the permissible $u \in \mathcal{U}$:

$$ {z_{\mathrm{d,min}}}^{(n)}(t) \leq {z_{\mathrm{d}}}^{(n)}(t) \leq {z_{\mathrm{d,max}}}^{(n)}(t) \tag{5.4a} $$

[1] This is the case because a flatness-based inverse is both a left and a right inverse [81] with respect to the flat output z.

with

$$
\begin{aligned}
{z_{\mathrm{d,min}}}^{(n)}(t) &= \min_{u\in\mathcal{U}} \left(\psi_{\mathrm{u}}^{-1}(z_{\mathrm{d}}(t),\ldots,{z_{\mathrm{d}}}^{(n-1)}(t),u)\right)\\
{z_{\mathrm{d,max}}}^{(n)}(t) &= \max_{u\in\mathcal{U}} \left(\psi_{\mathrm{u}}^{-1}(z_{\mathrm{d}}(t),\ldots,{z_{\mathrm{d}}}^{(n-1)}(t),u)\right)
\end{aligned}
\tag{5.4b}
$$

This limitation can be mapped to a limitation on the highest derivative of the IMC filter output $\tilde{y}_{\mathrm{d}}^{(r)}$ using Eq. (4.10b) which gives the desired permissible range

$$
\tilde{y}_{\mathrm{d,min}}^{(r)} \leq \tilde{y}_{\mathrm{d}}^{(r)} \leq \tilde{y}_{\mathrm{d,max}}^{(r)}. \tag{5.5}
$$

With these boundaries, the nonlinear IMC filter shown in Fig. 5.2 can be employed. Figure 5.4 shows the structure of a flatness-based IMC enhanced to respect input constraints.

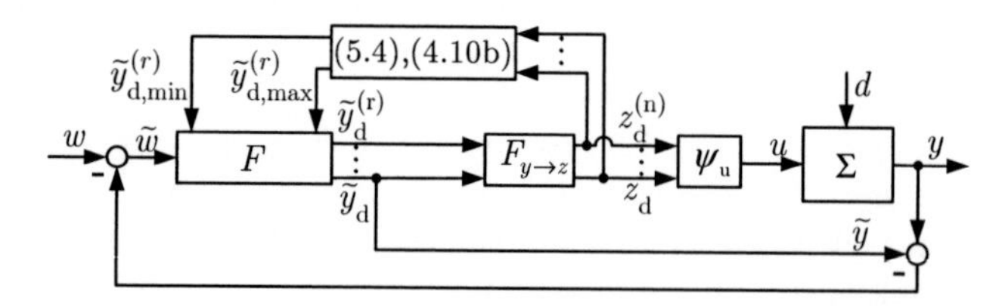

Fig. 5.4: Structure of a flatness-based IMC that respects input constraints.

A flatness-based IMC, as defined in Theorem 4.1, can be extended to take input constraints into account.

> **Corollary 5.1 (Respecting input constraints using a flatness-based IMC).** *The IMC controller Q is defined by Eqns.* (4.10a)-(4.10c) *and a limitation on the filter output* $\tilde{y}_{\mathrm{d}}$
>
> $$\tilde{y}_{\mathrm{d,min}}^{(r)}(t) \leq \tilde{y}_{\mathrm{d}}^{(r)}(t) \leq \tilde{y}_{\mathrm{d,max}}^{(r)}(t),$$
>
> *obtained from Eq.* (5.4) *inserted into Eq.* (4.10b).

The following example employs the proposed handling of input constraints. A linear plant is chosen for the sake of simplicity. The nonlinear case will be discussed in Part II for both automotive examples.

Example 5.1 (Respecting input constraints using a flatness-based IMC):
Consider the problem of designing an IMC controller (feedforward controller) Q for the model

$$\widetilde{\Sigma}(s) = \frac{s/5+1}{s^3+2s^2+s+1}. \tag{5.6}$$

The model can be represented in control normal form in state-space by

$$\begin{aligned} \dot{x} &= \begin{bmatrix} 0 & 1 & 0 \\ 0 & 0 & 1 \\ -1 & -1 & -2 \end{bmatrix} x + \begin{bmatrix} 0 \\ 0 \\ 1 \end{bmatrix} u \\ \tilde{y} &= \begin{bmatrix} 1 & \frac{1}{5} & 0 \end{bmatrix} x. \end{aligned} \tag{5.7}$$

All initial conditions are assumed to be zero. A linear system in control normal form always has a flat output $z = x_1$ [ii]:

$$\begin{aligned} z &= x_1 & \dot{z} &= \dot{x}_1 = x_2 \\ \ddot{z} &= \dot{x}_2 = x_3 & \dddot{z} &= \dot{x}_3 = -x_1 - x_2 - 2x_3 + u \\ & & &= -z - \dot{z} - 2\ddot{z} + u \end{aligned} \tag{5.8}$$

From Eq. (5.8) one finds the flatness-based feedforward control law (cf. Eq. (4.10c))

$$u = \psi_{\mathrm{u}}(\dddot{z}_{\mathrm{d}}, \ddot{z}_{\mathrm{d}}, \dot{z}_{\mathrm{d}}, z_{\mathrm{d}}) = \dddot{z}_{\mathrm{d}} + 2\ddot{z}_{\mathrm{d}} + \dot{z}_{\mathrm{d}} + z_{\mathrm{d}}. \tag{5.9}$$

In order to obtain the relationship $F_{\mathrm{y}\to\mathrm{z}}$ (cf. Eq. (4.10b)) between the desired model output $\tilde{y}_{\mathrm{d}}$ and the desired flat output z_{d} the output equation (see Eq. (5.7)) $\tilde{y} = x_1 + \frac{1}{5}x_2$ is used with $\tilde{y}_{\mathrm{d}} \stackrel{!}{=} \tilde{y}$ and the relationship (5.8) and yields

$$\tilde{y}_{\mathrm{d}} = z_{\mathrm{d}} + \dot{z}_{\mathrm{d}}/5 \qquad \dot{\tilde{y}}_{\mathrm{d}} = \dot{z} + \ddot{z}/5 \qquad \ddot{\tilde{y}}_{\mathrm{d}} = \ddot{z}_{\mathrm{d}} + \dddot{z}_{\mathrm{d}}/5. \tag{5.10}$$

The differential equation (5.10) needs to be solved (note that the signals $\tilde{y}_{\mathrm{d}}, \dot{\tilde{y}}_{\mathrm{d}}, \ddot{\tilde{y}}_{\mathrm{d}}$ are given from the IMC filter F) for $z_{\mathrm{d}}, \dot{z}_{\mathrm{d}}, \ddot{z}_{\mathrm{d}}$ and $\dddot{z}_{\mathrm{d}}$ with initial conditions $z_{\mathrm{d}}(0) = \dot{z}_{\mathrm{d}}(0) = \ddot{z}_{\mathrm{d}}(0) = 0$. The operator $F_{\mathrm{y}\to\mathrm{z}}$ presents this solution.

The IMC filter F is designed to exhibit the linear I/O behaviour (when not in saturation)

$$F(s) = \frac{1}{(s/\lambda+1)^2} = \frac{1}{s^2/\lambda^2 + 2s/\lambda + 1} \tag{5.11}$$

and implemented as SVF, as shown in Fig. 5.2. The dynamics of the filter F are in time domain (cf. (4.10a))

$$\tilde{w} = \frac{1}{\lambda^2}\ddot{\tilde{y}}_{\mathrm{d}} + \frac{2}{\lambda}\dot{\tilde{y}}_{\mathrm{d}} + \tilde{y}_{\mathrm{d}} \tag{5.12}$$

with initial conditions $\tilde{y}_{\mathrm{d}} = \dot{\tilde{y}}_{\mathrm{d}} = 0$. At this point in the example, a flatness-based IMC has been designed and it consists of Eq. (5.9), the solution of the differential equation (5.10), and the filter dynamics (5.12).

Now, input constraints are introduced as in Eq. (5.1). The following demonstrates how the restrictions on the highest derivative $\tilde{y}_{\mathrm{d}}^{(r)}$ of the filter can be obtained. From Eq. (5.9) one finds

$$\begin{aligned} \dddot{z}_{\mathrm{d,min}} &= \min_{u \in \mathcal{U}} \psi_{\mathrm{u}}^{-1}(z_{\mathrm{d}}, \dot{z}_{\mathrm{d}}, \ddot{z}_{\mathrm{d}}, u) = -2\ddot{z}_{\mathrm{d}} - \dot{z}_{\mathrm{d}} - z_{\mathrm{d}} + u_{\mathrm{min}} \\ \dddot{z}_{\mathrm{d,max}} &= \max_{u \in \mathcal{U}} \psi_{\mathrm{u}}^{-1}(z_{\mathrm{d}}, \dot{z}_{\mathrm{d}}, \ddot{z}_{\mathrm{d}}, u) = -2\ddot{z}_{\mathrm{d}} - \dot{z}_{\mathrm{d}} - z_{\mathrm{d}} + u_{\mathrm{max}}, \end{aligned} \tag{5.13}$$

the permissible range $\dddot{z}_{\mathrm{d,min}} \leq z_{\mathrm{d}} \leq \dddot{z}_{\mathrm{d,max}}$ of the highest derivative of the trajectory of the flat output z_{d}. Inserting this result into Eq. (5.10) one finds the desired limits

$$\begin{aligned} \ddot{\tilde{y}}_{\mathrm{d,min}} &= \ddot{z}_{\mathrm{d}} + \frac{1}{5} \dddot{z}_{\mathrm{d,min}} \\ \ddot{\tilde{y}}_{\mathrm{d,max}} &= \ddot{z}_{\mathrm{d}} + \frac{1}{5} \dddot{z}_{\mathrm{d,max}}. \end{aligned} \tag{5.14}$$

Hence, the flatness-based IMC will never violate the input constraints if the relationships (5.13) and (5.14) are used to limit the highest attainable derivative $\ddot{\tilde{y}}_{\mathrm{d}}$ of the IMC filter F, given in the differential equation (5.12) by $\tilde{y}_{\mathrm{d,min}} \leq \ddot{\tilde{y}}_{\mathrm{d}} \leq \tilde{y}_{\mathrm{d,max}}$.

Figure 5.5 shows the simulation results of the closed-loop system with a filter design using $\lambda = 4$ and the input constraints $u_{\mathrm{max}} = 2$ and $u_{\mathrm{min}} = -2$ for a step in the reference signal $w(t)$ occurring at time $t = 1$s. The dashed lines in Fig. 5.5 show the model output $\tilde{y}$ and input u (assuming an exact model) without any limitations on the input u. It is plotted to show how the right inverse $\widetilde{\Sigma}^{\mathrm{r}}$ calculates the input signal u without limitations. The plant model $\widetilde{\Sigma}$ has two weakly damped poles and, hence, tends to oscillate. During time $t = 1$s until time $t \approx 1.8$s, the dashed input u peaks to about $\max_t(u(t)) = 80$ and excites the model. In between time $t = 1.8$s and $t \approx 2$s the unlimited right inverse $\widetilde{\Sigma}^{\mathrm{r}}$ decelerates the plant model $\widetilde{\Sigma}$ with $\min_t(u(t)) \approx -11$ such that no oscillations occur and the plant output follows the filter output exactly $\tilde{y} = \tilde{y}_{\mathrm{d}}$.

If limitations on the input are active and the classical method of respecting input constraints, as discussed in Section 2.2.3, is used by limiting both plant and model inputs (cf. Fig. 2.12 on page 36), then the result is unsatisfactory (solid line in Fig. 5.5). In this case, the right inverse $\widetilde{\Sigma}^{\mathrm{r}}$ and the filter F do not know that the inputs u are limited. Hence, the input trajectory u (solid line) is identical to the unlimited case (dashed line) with the exception that it is cut off at $+2$ and -2. Thus, the controller Q is unaware that the dynamics of its output u do not reach the model and the plant. Note that the

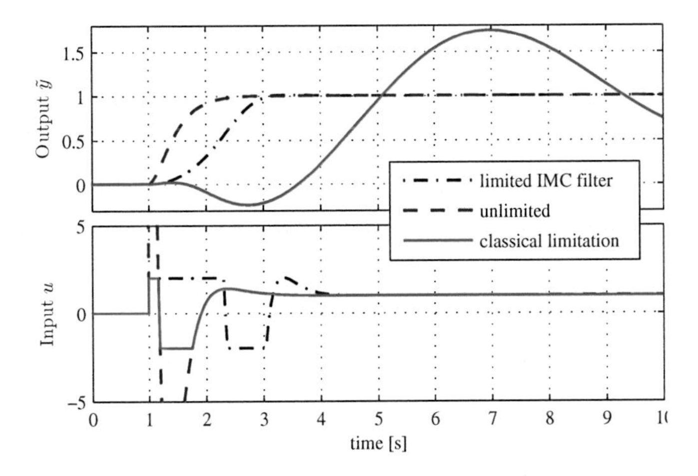

Fig. 5.5: Simulation results of Example 5.1, with $\lambda = 4$ and $u_{\max} = 2$ and $u_{\min} = -2$.

model output $\tilde{y}$ even first moves into the negative. Moreover, the model has been excited without actually compensating the weakly damped dynamics and, thus, oscillations occur and settle not until time $t \approx 40$s.

The result using the proposed limited IMC filter F (dash-dotted lines) is excellent. In this case, the IMC filter accounts for the limitation of the inputs. The limited IMC filter leads to an acceleration of the plant at the upper boundary $u(t) = u_{\max}$ for a much longer time period, namely $1s \leq t \leq 2.3$s. The deceleration occurs much later and also slightly longer than in the unlimited case from $2.3s \leq t \leq 3.2$s. Hence, the proposed limitation of the IMC filter F does not merely cut off the input trajectory. Something much more sophisticated happens: It produces an alternative trajectory $\tilde{y}_{\mathrm{d}}$ (dash-dotted line) which can be achieved by the model (i. e., $\tilde{y} = \tilde{y}_{\mathrm{d}}$) with permissible inputs $u \in \mathcal{U}$. Hence, the weakly damped dynamics are compensated by the right inverse $\widetilde{\Sigma}^{\mathrm{r}}$ since all of the calculated input u reaches the plant model $\widetilde{\Sigma}$. Moreover, all calculations are performed online since only differential equations need to be integrated. Thus, arbitrary reference signals $w(t)$ are handled directly without changing the design. ■

This method for treating input constraints produces an appealing result and it does not need to be designed by trial-and-error (i. e., there is no performance criterion to be chosen or poles to be placed) – it incorporates the knowledge of input constraints automatically into the existing design.

Input constraint with an IMC for systems in I/O normal form. For systems in I/O normal form, the relationship between the highest derivative $\tilde{y}_{\mathrm{d}}^{(r)}$ and the input u is given in Theorem 4.2 by Eq. (4.25c) and can be solved for $\tilde{y}_{\mathrm{d}}^{(r)}(t)$

$$\tilde{y}_{\mathrm{d}}^{(r)} = b(\boldsymbol{x}^*)u + a(\boldsymbol{x}^*).$$

Now, the desired limits on the highest derivative $\tilde{y}_{\mathrm{d}}^{(r)}$ (cf. Eq. (5.2b)) can be obtained directly by

$$\begin{aligned} \tilde{y}_{\mathrm{d,min}}^{(r)}(t) &= a(\boldsymbol{x}^*(t)) + \min\left(b(\boldsymbol{x}^*(t))u_{\min}, b(\boldsymbol{x}^*(t))u_{\max}\right) \\ \tilde{y}_{\mathrm{d,max}}^{(r)}(t) &= a(\boldsymbol{x}^*(t)) + \max\left(b(\boldsymbol{x}^*(t))u_{\min}, b(\boldsymbol{x}^*(t))u_{\max}\right). \end{aligned} \tag{5.15}$$

The necessary transformed state vector $\boldsymbol{x}^*$ in Eq. (5.15) is given partially by the IMC filter F itself (see Eq. (4.25a)) and the other part is the solution of the internal dynamics (see Eq. (4.25b)). Thus, the time-varying permissible range on the highest derivative $\tilde{y}_{\mathrm{d}}(t)^{(r)}$ in Eq. (5.15) can be computed.

An IMC for input-affine systems, as defined in Theorem 4.2 can be extended to respect input constraints Eq. (5.1).

Corollary 5.2 (IMC with input constraints for input-affine systems). *The IMC controller Q is then defined by Eqns.* (4.25a)-(4.25c) *and a limitation on the filter output $\tilde{y}_{\mathrm{d}}$*

$$\tilde{y}_{\mathrm{d,min}}^{(r)}(t) \leq \tilde{y}_{\mathrm{d}}^{(r)}(t) \leq \tilde{y}_{\mathrm{d,max}}^{(r)}(t),$$

obtained from Eq. (5.15).

Example 5.2 (IMC with input constraints using for input-affine systems):
Consider, again, the plant (5.7) from the previous example. Differentiating the output $\tilde{y}$ with respect to time gives

$$\tilde{y} = h(\boldsymbol{x}) = x_1 + \frac{1}{5}x_2 \qquad = \phi_1(\boldsymbol{x}) \tag{5.16a}$$

$$\dot{\tilde{y}} = L_f h(\boldsymbol{x}) = x_2 + \frac{1}{5}x_3 \qquad = \phi_2(\boldsymbol{x}) \tag{5.16b}$$

$$\ddot{\tilde{y}} = L_f^2 h(\boldsymbol{x}) = \frac{3}{5}x_3 - \frac{1}{5}(x_1 + x_2 + u) \tag{5.16c}$$

a relative degree $r = 2$. Choosing $\phi_3(\boldsymbol{x}) = x_1$ gives $\text{Rank}\,(\partial\boldsymbol{\phi}(\boldsymbol{x})/\partial\boldsymbol{x}) = 3$ and hence $\boldsymbol{\phi}(\boldsymbol{x}) = [\phi_1(\boldsymbol{x})\ \phi_2(\boldsymbol{x})\ \phi_2(\boldsymbol{x})]^T$ is a feasible transformation $\boldsymbol{x}^* = \boldsymbol{\phi}(\boldsymbol{x})$. The inverse transformation is

$$\boldsymbol{x} = \boldsymbol{\phi}^{-1}(\boldsymbol{x}^*) = \begin{bmatrix} x_3^* \\ 5(x_1^* - x_3^*) \\ 5x_2^* - 25(x_1^* - x_3^*) \end{bmatrix} \tag{5.17}$$

and gives the model $\widetilde{\Sigma}$ in the new coordinates $\boldsymbol{x}^*$ by $\dot{\boldsymbol{x}}^* = L_f\boldsymbol{\phi}(\boldsymbol{x})$ with the substitution $\boldsymbol{x} \mapsto \boldsymbol{\phi}^{-1}(\boldsymbol{x}^*)$

$$\begin{aligned} \dot{\boldsymbol{x}}^* &= \begin{bmatrix} x_2^* \\ \frac{1}{5}(-80x_1^* + 15x_2^* + 79x_3^*) + \frac{1}{5}u \\ 5(x_1^* - x_3^*) \end{bmatrix} \\ \tilde{y} &= x_1^*. \end{aligned} \tag{5.18}$$

With $x_1^* = \tilde{y}_\mathrm{d}$ and $x_2^* = \dot{\tilde{y}}_\mathrm{d}$ one finds from the above

$$u = 5\ddot{\tilde{y}}_\mathrm{d} - 10\dot{\tilde{y}}_\mathrm{d} + 80\tilde{y}_\mathrm{d} - 79x_3^* \tag{5.19}$$

with the internal dynamics

$$\dot{x}_3^* = 5(\tilde{y}_\mathrm{d} - x_3^*). \tag{5.20}$$

Thus, the resulting nonlinear IMC controller Q gets the signals $\tilde{y}_\mathrm{d}, \dot{\tilde{y}}_\mathrm{d}$ and $\ddot{\tilde{y}}_\mathrm{d}$ from the IMC filter dynamics (5.12), uses Eq. (5.19) to calculate the control input u and, for this purpose, has to solve the differential equation (5.20) of the internal dynamics for the state x_3^*. Again, it is proposed to implement the filter F as shown in Fig. 5.2.

At this point in the example, a nonlinear IMC has been developed, based on the I/O normal form, and does not respect input constraints. Now, consider the input constraints (5.1). From Eq. (5.19) one finds

$$\ddot{\tilde{y}}_{\mathrm{d,min}} = \frac{1}{5}(+10\dot{\tilde{y}}_\mathrm{d} - 80\tilde{y}_\mathrm{d} + 79x_3^*) + \frac{1}{5}u_{\min} \tag{5.21}$$

$$\ddot{\tilde{y}}_{\mathrm{d,max}} = \frac{1}{5}(+10\dot{\tilde{y}}_\mathrm{d} - 80\tilde{y}_\mathrm{d} + 79x_3^*) + \frac{1}{5}u_{\max} \tag{5.22}$$

as the desired limitations on the highest derivative of the IMC filter. The simulation results are identical to the ones shown in Example 5.1, since the same model is controlled. ■

5.2 IMC for Systems with Ill-Defined Relative Degree

The relative degree r is said to be ill-defined if it changes its value in dependence on the current state $\boldsymbol{x}(t)$, say at $\bar{\boldsymbol{x}}(t)$. A right inverse $\widetilde{\Sigma}^{\mathrm{r}}$, designed for the relative degree r_0 found at some $\boldsymbol{x}(t)$, becomes singular at the state $\bar{\boldsymbol{x}}(t)$ (where the relative degree is $> r_0$), in the sense that the output u of the inverse $\widetilde{\Sigma}^{\mathrm{r}}$ is not defined (i. e., $u \notin \mathcal{L}_\infty$).

The typical approach of dealing with a finite raise of the relative degree r at the state $\bar{\boldsymbol{x}}(t)$ has been discussed by e. g., [48, 51]. Their approach includes generating k further differentiations of the output $\tilde{y}$ until the relationship between the highest derivative of the input $u^{(k)}$ and the highest derivative of the output $\tilde{y}^{(r_0+k)}$ is not singular if solved for $u^{(k)}$. Then, $u^{(k)}$ is chosen as the new input $\nu = u^{(k)}$ and its lower derivatives follow from an integrator chain. This can be interpreted as an extension of the state-space of the model. In light of Corollary 3.1, such a procedure results in a smoother u as a larger constant relative degree r is enforced globally and, thus, a smoother I/O behaviour of the controlled system follows.

If this strategy is successful, then this effectively yields a non-singular inverse that is valid for the whole state trajectory $\boldsymbol{x}$. This right inverse maps a given (smoother) trajectory $\tilde{y}_{\mathrm{d}}^{(r_0+k)}$ into the new input ν. This approach can also be followed for the proposed design of an IMC. However, if k does not exist, then the approach fails to generate a feasible right inverse. The following example illustrates this.

Example 5.3 (Model with a Singularity):
Consider the system

$$\begin{aligned} \dot{x}_1 &= x_2 \\ \dot{x}_2 &= -x_2 - 4x_1 + (x_2 - 2)u \\ \tilde{y} &= x_1 \end{aligned} \tag{5.23}$$

with $x_1(0) = x_2(0) = 0$, for which one finds

$$\begin{aligned} \dot{\tilde{y}} &= x_2 \\ \ddot{\tilde{y}} &= \dot{x}_2 = -x_2 - 4x_1 + (x_2 - 2)u. \end{aligned} \tag{5.24}$$

This system has a relative degree of $r = 2$ for $x_2(t) \neq 2$. Since any arbitrary number of further differentiations of the output $\tilde{y}$ does not establish a relation to the input u for $x_2(t) = 2$, the system has no relative degree (i. e., $r \rightarrow \infty$) for $x_2(t) = 2$.

The right inverse $\widetilde{\Sigma}^{\mathrm{r}}$ is established using the second derivative $\ddot{\tilde{y}}$ as

$$u = \frac{4\tilde{y} + \dot{\tilde{y}} + \ddot{\tilde{y}}}{-2 + \dot{\tilde{y}}} \tag{5.25}$$

and it becomes singular if the relative degree is lost:

$$\lim_{\tilde{y}_{\mathrm{d}}(t) \rightarrow 2} u(t) = \infty. \tag{5.26}$$

■

This section proposes a method which deals with a loss in relative degree and, thus, presents an IMC which deals with singular model inverses.

Main idea. As in the previous section, where input constraints were discussed, it is proposed to alter the filter F such that it generates a trajectory $\tilde{y}_{\mathrm{d}}$ that can be reached by the model output $\tilde{y}$ exactly (i. e., $\tilde{y} = \tilde{y}_{\mathrm{d}}$), with permissible inputs $u \in \mathcal{U}$.

Remark 5.1. It is important to realise that a loss in relative degree $r \rightarrow \infty$ is a system property and stems from the function of the plant. It results in an autonomous behaviour of the plant model since the input u has no effect on the model output $\tilde{y}$. Hence, at the singularity $\bar{x}(t)$, it is impossible to find a right inverse $\widetilde{\Sigma}^{\mathrm{r}}$ that is defined for an arbitrarily demanded trajectory $\tilde{y}_{\mathrm{d}}$.

While the model is in the singularity, the only possibility to deal with an ill-defined relative degree is to alter the IMC filter F such that it produces a demanded trajectory $\tilde{y}_{\mathrm{d}}$ that coincides with the autonomous output $\tilde{y}$ of the model. Then, the right inverse is still not defined, but the input u can be selected arbitrarily (the right inverse becomes a relation with an infinite number of possible outputs $u(t) \in \mathbb{R}$). Thus, the input $u(t)$ can be selected to be admissible $u \in \mathcal{U}$.

Therefore, the goal becomes finding such an IMC filter F.

The solution to singular inverses is to include input limitations:

Theorem 5.2 (Handling Singular Model Inverses with IMC). *If input constraints using any boundaries $u_{\min} \leq u_{\max}$ are treated as proposed in Theorem 5.1, then the output u of the right inverse $\widetilde{\Sigma}^{\mathrm{r}}$ can* always *be selected to be in the permissible range $u \in \mathcal{U}$ and the model output $\tilde{y}$ is equal to the filter output (i. e., $\tilde{y}_{\mathrm{d}} = \tilde{y}$).*

Proof. The function $\varphi(\boldsymbol{x}, u)$ exists for all possible trajectories $\boldsymbol{x}$ (and u), independent of existence or the current value of r (i. e., if r raises its value, then φ becomes independent on u, but it is still defined for the relative degree r_0 that it was originally designed for.)

The method of respecting input constraints, described in Theorem 5.1, ensures (see (5.2a) and (5.2b)) that, at time t, the highest derivative $\tilde{y}_{\mathrm{d}}^{(r_0)}$ is chosen such that

$$\min_{u \in \mathcal{U}} \varphi(\boldsymbol{x}(t), u) \leq \tilde{y}_{\mathrm{d}}^{(r_0)} \leq \max_{u \in \mathcal{U}} \varphi(\boldsymbol{x}(t), u) \tag{5.27}$$

holds. This holds at every instance in time $t \in \mathcal{T}$.

The function $\varphi(\boldsymbol{x}, u)$ exists also if the relative degree $r \to \infty$ vanishes. Then, $\varphi(\boldsymbol{x}, u)$ is only dependent on $\boldsymbol{x}$ (i. e., $\varphi(\boldsymbol{x}, u) = \varphi(\boldsymbol{x}, \cdot)$). Therefore, the limits $\min_{u \in \mathcal{U}} \varphi(\boldsymbol{x}(t), u)$ and $\max_{u \in \mathcal{U}} \varphi(\boldsymbol{x}(t), u)$ produce the same value and the highest derivative of the filter output $\tilde{y}_{\mathrm{d}}^{(r_0)}$ is chosen to be identical to the autonomous behaviour of the model output (i. e., $\tilde{y}_{\mathrm{d}}^{(r_0)} = \tilde{y}^{(r_0)} = \varphi(\boldsymbol{x}, \cdot)$). This demand can be met with any $u \in \mathcal{U}$. □

Interpretation. Loosely speaking, the above Theorem states that, since the input u is limited, it cannot be chosen as infinite by the right inverse. As a result, it follows that the right inverse produces a feasible output $u \in \mathcal{U}$. However, this method has a substantial demerit: It forces the system into an autonomous behaviour even if the relative degree changes to a larger but finite value. In such a case, the system may not recover from its self-induced autonomy despite the ability of the model and the plant to leave this state.

Literature. Several publications have addressed the issue of singularities that originate from an ill-defined relative degree where the relative degree vanishes completely (e. g., [13, 34, 49, 62, 68]). All publications address changing the inverse when the plant is close to the singularity. Thus, their approach is inherently different from the one portrayed here: Here, it is not proposed to change the inverse. Rather, the trajectory is (automatically) chosen to avoid the singularities in the sense that a permissible $u \in \mathcal{U}$ can be chosen at any instance in time. The inverse itself is not changed. Hence, while the approach proposed in this thesis guarantees exact tracking of an altered trajectory[2], the approaches in the cited

[2] The altered trajectory is the only possible trajectory in the case of a vanishing r. Thus, the choice of the trajectory is not to be considered arbitrary.

literature do not change the trajectory, but allow for imperfect tracking while travelling near a singularity.

The advantage of the approach proposed here is that once input constraints are considered (which they virtually always have to be) singularities are also treated in the sense that the input signal u is non-explosive (i. e., $u \in \mathcal{L}_\infty$). Thus, *singularities of model inverses do not necessitate a special treatment.*

Example 5.4 (Example 5.3 continued):
In order to design an IMC controller Q for the model (5.23) presented in the previous example, input constraints need to be assumed

$$u_{\min} \le u(t) \le u_{\max} \tag{5.28}$$

with $u_{\max} = -u_{\min} = 1000$. The value of 1000 is chosen arbitrarily. From Eq. (5.25) one determines the boundaries for the limited IMC filter F as

$$\begin{aligned} \ddot{\tilde{y}}_{\mathrm{d,min}} &= -4\tilde{y}_{\mathrm{d}} - \dot{\tilde{y}}_{\mathrm{d}} + \min\left((\dot{\tilde{y}}_{\mathrm{d}} - 2)u_{\min}, (\dot{\tilde{y}}_{\mathrm{d}} - 2)u_{\max}\right) \\ \ddot{\tilde{y}}_{\mathrm{d,max}} &= -4\tilde{y}_{\mathrm{d}} - \dot{\tilde{y}}_{\mathrm{d}} + \max\left((\dot{\tilde{y}}_{\mathrm{d}} - 2)u_{\min}, (\dot{\tilde{y}}_{\mathrm{d}} - 2)u_{\max}\right). \end{aligned} \tag{5.29}$$

Thus, an IMC controller Q for the plant model (5.23) consists of the equation (5.25) of the right inverse and the IMC filter dynamics $\tilde{w} = \ddot{\tilde{y}}_{\mathrm{d}}/\lambda^2 + 2\dot{\tilde{y}}_{\mathrm{d}}/\lambda + \tilde{y}_{\mathrm{d}}$ (with zero initial conditions) that is restricted by $\ddot{\tilde{y}}_{\mathrm{d,min}} \le \ddot{\tilde{y}}_{\mathrm{d}} \le \ddot{\tilde{y}}_{\mathrm{d,max}}$ from Eq. (5.29).

Figure 5.6 shows the simulation results of this example for a reference step $w(t) = 0$ for $t < 1\mathrm{s}$ and $w(t) = 10$ for $t \ge 1\mathrm{s}$. Without any singularity handling the input u would become singular.

Figure 5.6 shows that the input u does not exceed the lower limit -1000. The derivative of the output $\dot{\tilde{y}}_{\mathrm{d}}$ does not move into the singularity at $\dot{\tilde{y}}_{\mathrm{d}} = x_2 = 2$ since this is prevented by the introduction of the input constraints. Note that the introduced system is theoretically unable to go past the singularity if u is to be defined at all times. ■

5.3 *Non-Minimum Phase IMC Design using a Perfect Inverse*

If the inverse of a model is unstable then the model is NMP (see Definition 3.6). It follows from Property 3.1 that the nonlinear IMC design from Algorithm 3.1 (using Eq. (3.28)) leads to an internally unstable closed-loop if the model $\tilde{\Sigma}$ is NMP.

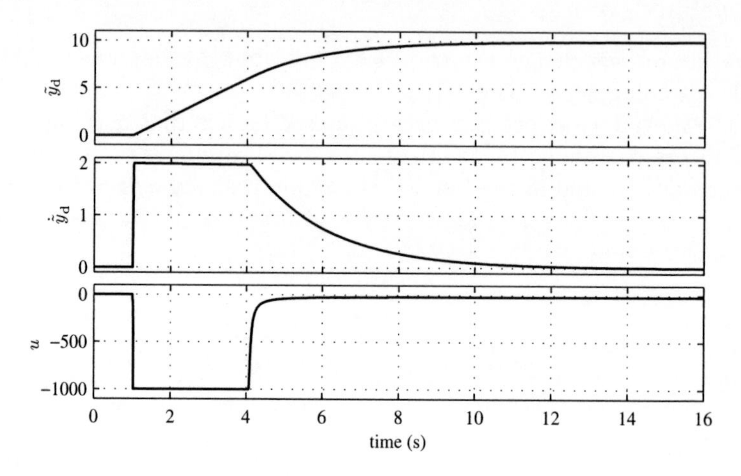

Fig. 5.6: Simulation results with singularity handling of Example 5.4. The IMC filter uses $\lambda = 1$.

This section proposes a novel method to design an IMC for an NMP model $\widetilde{\Sigma}$. As a by-product, the proposed method allows to develop a feed-forward controller for nonlinear NMP systems which can be computed on-line. The ability to control a nonlinear NMP system is a major advantage of the IMC structure over feedback linearisation methods.

It is proposed to use an exact inverse $\widetilde{\Sigma}^{\mathrm{r}}$ of the NMP plant model $\widetilde{\Sigma}$. In order to obtain a feasible controller Q, two subsequent steps are suggested: The first step requires to introduce the NMP behaviour of the model $\widetilde{\Sigma}$ into the IMC filter F. This leads to an internally unstable zero-pole cancellation, but stable I/O behaviour of $\widetilde{\Sigma}^{\mathrm{r}} \circ F$. In the second step the unstable internal dynamics of the inverse $\widetilde{\Sigma}^{\mathrm{r}}$ are removed and replaced by the solution of the filter, resulting in an internally stable IMC controller.

Before the method is introduced in detail, it is briefly discussed why inverses of NMP stable models become unstable.

5.3.1 Background

In order to explain the internal behaviour of an NMP model, the input affine system (4.15) is used. The internal dynamics of (4.15) become visible when the system is transformed into the I/O normal form (4.19). Figure 5.7 is a representation of the I/O normal form. Note that in Fig. 5.7,

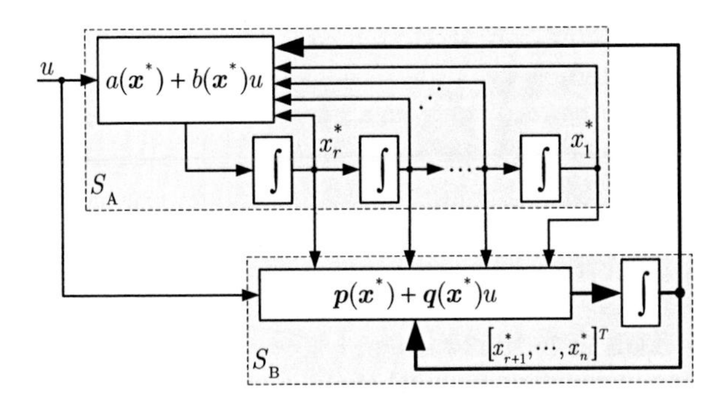

Fig. 5.7: Structure of an input-affine system $\widetilde{\Sigma}$ in I/O normal form.

the system has been structured into the subsystems, S_{A} and S_{B}. The subsystem S_{A} contains the input relationship (4.19b) and the integrator chain of length r. Subsystem S_{B} represents the internal dynamics (4.19c).

The systems S_{A} and S_{B} form a feedback connection. The connection signals are the states $x_1^*, \ldots, x_r^*$ from S_{A} into S_{B} and the states $x_{r+1}^*, \ldots, x_n^*$ of the internal dynamics (4.19c) from S_{B} into S_{A}. Since only stable systems are considered, this feedback connection is stable by definition.

In the case of an NMP system $\widetilde{\Sigma}$, one or both of the subsystems S_{A} and S_{B} are open-loop unstable. In this case, the subsystems stabilise each other[3]. When constructing the right inverse $\widetilde{\Sigma}^{\mathrm{r}}$, the input u is chosen as in Eq. (4.22), in order for the subsystem S_{A} to behave like a pure integrator chain of length r. As an undesired result, it cancels the feedback of the states of the internal dynamics S_{B} on S_{A}. Moreover, a new feedback loop is closed via the input calculation (4.22). This feedback, however, may not stabilise the internal dynamics.

In summary, instability of an NMP right inverse $\widetilde{\Sigma}$ occurs for one of two reasons:

1. The subsystem S_{B} is unstable. Due to cancellation of the stabilising feedback signal, S_{B} remains unstable.
2. It is the introduction of the input u by (4.22) which, inserted into S_{B}, renders the internal dynamics unstable.

[3] For example: Assume S_{A} is stable. Then S_{A} acts as a stabilising state feedback controller for the unstable internal dynamics S_{B}.

Literature. No literature was found on the topic of NMP nonlinear IMC design. However, since the IMC design is essentially based on the construction of a feedforward controller, some literature on nonlinear NMP feedforward control design is presented.

In [59] it is proposed, as in the linear case (cf. Section 2.2.2), to split the model $\widetilde{\Sigma}$ into a minimum phase part $\widetilde{\Sigma}_{\mathrm{MP}}$ and a nonlinear all-pass part $\widetilde{\Sigma}_{\mathrm{NMP}} = \widetilde{\Sigma}_{\mathrm{A}}$. Unfortunately, this problem has only been solved for input-affine models of up to second order. As in the linear case, the resulting right-inverse leads to an ISE-optimal controller. As it was established in Example 2.2 on page 34, that ISE-optimal controllers are of no concern for an automotive control engineer, the approach in [59] is not followed here. However, if a general solution to split a nonlinear model into the above mentioned parts can be found, it would be of great value, also for IMC design.

The work of [65] discusses how the internal model *principle*, which also holds for nonlinear systems, can be exploited for control design. The main problem solved is to follow a reference trajectory generated by a known exosystem or to attenuate disturbances generated by a known exosystem. As a result, also a non-minimum phase plant can be made to asymptotically follow the trajectory of an exosystem at the cost of some (unknown) transient tracking error.

The contributions [12, 23, 24] establish a feedforward controller for a priori known reference trajectories with compact support. The control scheme does not produce a transient error despite the non-minimum phase characteristic of the plant. This, however, comes at the cost of a non-causal inverse as the nonlinear and unstable zero dynamics of the plant need to be integrated backwards in time to obtain a feasible initial condition. Despite the shortcoming of a non-causal controller, the authors have established that the design of a feedforward controller for an a priori known trajectory can be interpreted as a two-point boundary value problem. The solution of this problem yields the necessary control input. This idea was picked up by [36] and generalised to trajectories without compact support. Additionally, [36] was able to put constraints on the model states and the control input during the transient behaviour of the trajectory. However, the solution to the trajectory tracking problem needs to be found off-line, in particular by numerical methods.

In conclusion, the solutions above are all interesting and feasible for their respective control problems. However, none solves the control problem discussed in this thesis as they all violate at least one of the following requirements (cf. Section 1.3.2):

- An exosystem is not given. Rather, the reference values $w(t)$ considered here are arbitrary $w \in \mathcal{L}_\infty$ as they are influenced directly by the driver.
- The reference trajectory w is not known before hand, which makes a non-causal inverse or an off-line solution to trajectory tracking impossible.

The problem at hand (stated in Chapter 3.5) appears to be much simpler than the control problems as discussed by the aforementioned publications. However, no solutions on this control problem were found. In the view of the available literature one can interpret the problem that, here, it is the exosystem to be designed (namely the IMC filter F) such that a causal inverse $\widetilde{\Sigma}^{\mathrm{r}}$ (the composition of the two is the IMC controller Q) tracks a reference signal w without producing an exploding control input u. This approach is persued in this chapter.

5.3.2 Introducing the Non-Minimum Phase IMC Design Using Linear Systems

In the following, a method to design an IMC for NMP models is presented which retains the IMC design philosophy in such that the IMC filter F is designed as the demanded I/O behaviour of the nominal closed-loop. This method is first introduced by the equivalent procedure for linear systems. Then, the necessary amendments for nonlinear systems are given, followed by a summary in a theorem.

Main idea. It is proposed to design the linear IMC controller $Q(s)$ for linear NMP models $\widetilde{\Sigma}$ as in the classical IMC design procedure (cf. Eq. (2.14a)), for linear minimum phase model, namely by

$$Q(s) = \widetilde{\Sigma}^{-1}(s)F(s).$$

Hence, the IMC controller $Q(s)$ is obtained by a perfect inversion of an NMP model. Figure 5.8 shows this open loop structure. Note that Fig. 5.8 displays the lower branch of the IMC structure given in Fig. 2.1 or Fig. 3.3.

If the model $\widetilde{\Sigma}(s)$ is NMP, it has at least one zero[4] z in the right half of the complex plane ($\mathrm{Re}\,(z) > 0$). The feedforward behaviour $\widetilde{\Sigma}(s)Q(s)$

[4] Note that pure time delays are not discussed in this work.

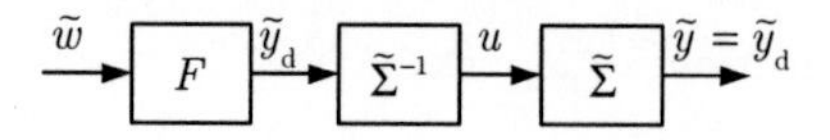

Fig. 5.8: Feedforward control structure.

is I/O stable if and only if the IMC filter $F(s)$ shares all right-half plane zeros of the model $\tilde{\Sigma}(s)$. The zeros of the filter are used to cancel the unstable poles that result from the inversion $\tilde{\Sigma}^{-1}(s)$. Note that an exact cancellation is, in principle, possible, since $\tilde{\Sigma}(s)$ is artificial and, thus, perfectly known.

Remark 5.2 (Internal stability)**.** Internal stability requires that all transfer functions from every possible input to every possible output must be stable. The expression "possible input" means that an exogenous signal is allowed to alter any signal of the control structure.

Considering the signals in Fig. 5.8, only $\tilde{w}$ can be affected by exogenous signals because *all* of the operators are artificial. This means that all intermediary signals from F to $\tilde{\Sigma}$ are not subject to alteration by exogenous signals, like disturbances. One could, therefore, argue that the feedforward control structure with controller $Q(s)$ and model $\tilde{\Sigma}(s)$ is internally stable because the systems $F(s)$, $\tilde{\Sigma}^{-1}(s)F(s)$, and $\tilde{\Sigma}(s)\tilde{\Sigma}^{-1}(s)F(s)$ are I/O stable.

Although this is true from a mathematical perspective, it does not hold true for the actual implementation. Since artificial systems are ultimately implemented on a digital processor, all signal values are restricted to a certain precision. Hence, there are numerical errors on all signals which render the structure Fig. 5.8 internally unstable because the transfer function $\tilde{\Sigma}^{-1}(s)$ is unstable.

Interpretation. Obtaining the IMC controller Q from an inversion of an NMP model results in an internally unstable feedforward path. However, it can be used to review some useful facts:

- There exist some input trajectories[5] $\tilde{y}_\mathrm{d} = F(s)\tilde{w}$ to the inverse $\tilde{\Sigma}^{-1}(s)$ which, despite of the instability of $\tilde{\Sigma}^{-1}(s)$, *do not* result in an exploding output signal u.

[5] These trajectories $\tilde{y}_\mathrm{d}$ do not excite the unstable modes of the model $\tilde{\Sigma}$.

- These input signals $\tilde{y}_\mathrm{d}$ can be obtained by including the right-half plane zeros of the model $\widetilde{\Sigma}(s)$ into the IMC filter $F(s)$.

The main benefit of this approach is that the IMC philosophy is retained for NMP systems. Thus, the IMC filter $F(s)$ is designed as the demanded I/O behaviour from $w(s)$ to $y(s)$ of the closed loop and

$$\frac{y(s)}{w(s)} = F(s) \tag{5.30}$$

holds in the nominal case for NMP systems. The deficiency of this approach is that F can no longer be selected as a linear low-pass filter without any zeros, as introduced in Chapter 3.

Internal stability. One can obtain an internally stable feedforward control structure by implementing the minimal realisation of $Q(s)$. Thus, the zero-pole cancellation of $\widetilde{\Sigma}^{-1}(s)F(s)$ is performed before the implementation of $Q(s)$. Comparing this procedure with the results of the linear IMC for NMP systems in Section 2.2.2 (therein, especially Eq. (2.26), together with Eq. (2.27)), one finds that they are equivalent.

The following brief example illustrates the considerations above.

Example 5.5 (NMP IMC design for linear systems.):
Consider the model $\widetilde{\Sigma}(s) = \frac{s-1}{(s+2)(s+4)}$ with an NMP zero at $z = 1$. The IMC filter is chosen as $F(s) = \frac{-(s-1)}{(s/\lambda+1)^2}$ and also exhibits the zero at $z = 1$. The model inverse is given by $\widetilde{\Sigma}^{-1}(s) = \frac{(s+2)(s+4)}{s-1}$ and, consequentially, has an unstable pole at $p = 1$. The IMC controller $Q(s)$ results in

$$\begin{aligned} Q(s) &= \widetilde{\Sigma}^{-1}(s)F(s) \\ &= \frac{(s+2)(s+4)}{s-1} \cdot \frac{s-1}{(s/\lambda+1)^2} \\ &= \frac{(s+2)(s+4)}{1} \cdot \frac{1}{(s/\lambda+1)^2} \end{aligned} \tag{5.31}$$

where the last line of the equation above presents the minimal realisation of $Q(s)$, which clearly is a stable transfer function due to the choice of $\lambda > 0$.
■

5.3.3 Non-Minimum Phase IMC Design for Nonlinear Systems

Main idea. It is proposed to extend the above design for NMP linear systems to nonlinear input-affine[6] systems. Thus, the idea is to construct an IMC controller Q for nonlinear NMP systems in the same manner as it is designed for MP systems in Eq. (3.28), namely by perfect inversion of the nonlinear NMP model

$$Q = \widetilde{\Sigma}^{\mathrm{r}} F.$$

Loosely speaking, the idea is to

1. design a stable IMC filter F, which exhibits the same NMP behaviour as the model $\widetilde{\Sigma}$, and
2. find an internally stable implementation of Q.

To begin with, the following defines the condition under which the trajectories of the internal dynamics of two systems are identical.

Definition 5.1 (Identical behaviour of the internal dynamics of two systems). Assume two input-affine systems A and B, both given in the I/O normal form as in Eq. (4.19), but with all variables defined with the subscript A or B, respectively. Then, the internal dynamics of A and B are identical if

$$\boldsymbol{p}_{\mathrm{A}}(\boldsymbol{x}_{\mathrm{A}}^{*}) + \boldsymbol{q}_{\mathrm{A}}(\boldsymbol{x}_{\mathrm{A}}^{*}) u_{\mathrm{A}} - \boldsymbol{p}_{\mathrm{B}}(\boldsymbol{x}_{\mathrm{B}}^{*}) - \boldsymbol{q}_{\mathrm{B}}(\boldsymbol{x}_{\mathrm{B}}^{*}) u_{\mathrm{B}} = \boldsymbol{0} \tag{5.32a}$$

holds with equal initial conditions

$$\begin{bmatrix} x_{\mathrm{A},r+1}^{*}(0) & \cdots & x_{\mathrm{A},n}^{*}(0) \end{bmatrix}^{T} = \begin{bmatrix} x_{\mathrm{B},r+1}^{*}(0) & \cdots & x_{\mathrm{B},n}^{*}(0) \end{bmatrix}^{T}. \tag{5.32b}$$

◇

Definition 5.1 establishes equality of two sets of differential equations by requesting identical right hand sides (see Eq. (5.32a)) and identical initial conditions (see Eq. (5.32b)). As a result, the trajectories of the internal dynamics of the two systems A and B are identical. Thus,

$$\begin{bmatrix} x_{\mathrm{A},r+1}^{*}(t) & \cdots & x_{\mathrm{A},n}^{*}(t) \end{bmatrix}^{T} = \begin{bmatrix} x_{\mathrm{B},r+1}^{*}(t) & \cdots & x_{\mathrm{B},n}^{*}(t) \end{bmatrix}^{T} \tag{5.33}$$

holds for each instance in time $t \geq 0$.

[6] The proposed design is valid for all nonlinear systems that can be implemented in an I/O normal form but it is introduced using this limited system class for reasons of clarity.

Remark 5.3. Definition 5.1 does not imply equality of the individual vector-valued functions $\boldsymbol{p}_{(\cdot)}$ or $\boldsymbol{q}_{(\cdot)}$ nor does it imply equality of the input signals $u_{(\cdot)}$. Hence, in general, the trajectories of the states of the internal dynamics of A and B may be identical even if

$$\begin{aligned} \boldsymbol{p}_{\mathrm{A}}(\boldsymbol{x}_{\mathrm{A}}^*) &\neq \boldsymbol{p}_{\mathrm{B}}(\boldsymbol{x}_{\mathrm{B}}^*) \\ \boldsymbol{q}_{\mathrm{A}}(\boldsymbol{x}_{\mathrm{A}}^*) &\neq \boldsymbol{q}_{\mathrm{B}}(\boldsymbol{x}_{\mathrm{B}}^*) \\ u_{\mathrm{A}} &\neq u_{\mathrm{B}} \end{aligned}$$

holds. This is exemplified here: The two systems $\dot{x}_{\mathrm{A}} = u_{\mathrm{A}}$ and $\dot{x}_{\mathrm{B}} = -3x_{\mathrm{B}} + u_{\mathrm{B}}^2$ have identical solutions $x_{\mathrm{A}} = x_{\mathrm{B}}$ with identical initial conditions $x_{\mathrm{A}}(0) = x_{\mathrm{B}}(0)$ and the input signal $u_{\mathrm{A}} = -3x_{\mathrm{B}} + u_{\mathrm{B}}^2$.

Moreover, Definition 5.1 does not make any implications about the states $x^*_{:,1}, \ldots, x^*_{:,r}$ which do not belong to the internal dynamics.

Using Definition 5.1 as a starting point, the following lemma gives the condition under which the internal dynamics of the IMC filter F are equal to the internal dynamics of the inverse $\widetilde{\Sigma}^{\mathrm{r}}$ assuming a series connection of F and $\widetilde{\Sigma}^{\mathrm{r}}$ as it appears in an IMC design (cf. Fig. 3.6). To this end, the model $\widetilde{\Sigma}$ is regarded under the influence of input u from Eq. (4.22).

Lemma 5.1 (Identical internal dynamics of IMC filter F and model $\widetilde{\Sigma}$ under the influence of the right inverse $\widetilde{\Sigma}^{\mathrm{r}}$). *Assume an input-affine model $\widetilde{\Sigma}$, given in I/O normal form as defined in Eq. (4.19), with input u, and output $\tilde{y} = x_1^*$. Further, assume an input-affine system F which is also given in I/O normal form, as defined in Eq. (4.19), but with all variables defined with a subscript* F*. Hence, the states of F are $x^*_{\mathrm{F},1}, \ldots, x^*_{\mathrm{F,n}}$ with input $u_{\mathrm{F}} = \tilde{w}$, the output is $\tilde{y}_{\mathrm{d}} = y_{\mathrm{F}} = x^*_{\mathrm{F},1}$ and $\tilde{y}_{\mathrm{d}}^{(r)} = \dot{x}_{\mathrm{F},r^*}$ holds. The system F is chosen to have the same number of states n and the same relative degree r as the system $\widetilde{\Sigma}$. Finally, let the input u to the model $\widetilde{\Sigma}$ be defined by Eq. (4.22), or, in a more general setting, by*

$$u = \frac{1}{b(\boldsymbol{x}^*)}\left(-a(\boldsymbol{x}^*) + \tilde{y}_{\mathrm{d}}^{(r)}\right) \triangleq \gamma\left(\boldsymbol{x}^*, \tilde{y}_{\mathrm{d}}^{(r)}\right). \tag{5.34}$$

Note that the input transformation of Eq. (5.34) is part of the right inverse $\widetilde{\Sigma}^{\mathrm{r}}$ of the model $\widetilde{\Sigma}$.

With the assumptions above, the trajectory of the states of the internal dynamics of the model $\widetilde{\Sigma}$ under the influence of the right inverse $\widetilde{\Sigma}^{\mathrm{r}}$ are *identical to the internal dynamics of the IMC filter F if and only if*

$$\boldsymbol{p}(\boldsymbol{x}^*) + \boldsymbol{q}(\boldsymbol{x}^*) \cdot \gamma\left(\boldsymbol{x}^*, \tilde{y}_{\mathrm{d}}^{(r)}\right) - \boldsymbol{p}_{\mathrm{F}}(\boldsymbol{x}_{\mathrm{F}}^*) - \boldsymbol{q}_{\mathrm{F}}(\boldsymbol{x}_{\mathrm{F}}^*)\tilde{w} = \boldsymbol{\delta}\left(\boldsymbol{x}^*, \boldsymbol{x}_{\mathrm{F}}^*\right) \quad (5.35a)$$

holds, where the function $\boldsymbol{\delta}\left(\boldsymbol{x}^*, \boldsymbol{x}_{\mathrm{F}}^*\right)$ has the property

$$\boldsymbol{\delta}\left(\boldsymbol{x}^*(t), \boldsymbol{x}_{\mathrm{F}}^*(t)\right) = \boldsymbol{0} \text{ for } \boldsymbol{x}^*(t) = \boldsymbol{x}_{\mathrm{F}}^*(t). \quad (5.35b)$$

Proof. Lemma 5.1 meets Definition 5.1 in such that Eq. (5.35a) corresponds to Eq. (5.32a) and, since the initial condition of IMC filter F and model $\widetilde{\Sigma}$ are demanded equal ($\boldsymbol{x}^*(0) = \boldsymbol{x}_{\mathrm{F}}^*(0)$) by the IMC design, the function $\boldsymbol{\delta}\left(\boldsymbol{x}^*, \boldsymbol{x}_{\mathrm{F}}^*\right)$ will be zero at $t = 0$ (which meets Eq. (5.32b)). Due to Eq. (5.35a), it will remain zero for each instance in time $t > 0$. □

Remark 5.4. Note that, if two systems A and B have identical zero dynamics, it does not necessarily follow that the behaviour of their internal dynamics are identical in the sense of Lemma 5.1 or Definition 5.1.

This holds, since the notion of zero dynamics assumes particular initial conditions and a particular output trajectory, which cannot be assumed to be given in the general case.

The following corollary establishes an important intermediary result and is the conclusion of Lemma 5.1 and the IMC design procedure (cf. Section 3.5). It is the basis for an internally stable implementation of the proposed IMC controller $Q = \widetilde{\Sigma}^{\mathrm{r}} F$.

Corollary 5.3 (Identical state trajectories in feedforward control). *Assume the model $\widetilde{\Sigma}$ and the filter F to be given as in Lemma 5.1. If the conditions of Lemma 5.1 hold true (i. e., the state trajectories of the internal dynamics of IMC filter F and right inverse $\widetilde{\Sigma}^{\mathrm{r}}$ are identical) then the trajectories of all states of F and $\widetilde{\Sigma}^{\mathrm{r}}$ are equal. Thus, the equality*

$$\left[x_1^*(t), \quad \cdots, \quad x_n^*(t)\right]^T = \left[x_{\mathrm{F},1}^*(t), \quad \cdots, \quad x_{\mathrm{F},n}^*(t)\right]^T \quad (5.36)$$

applies at each instance in time $t \geq 0$.

Since, for IMC design, the initial states of the Filter F and the right inverse $\widetilde{\Sigma}^{\mathrm{r}}$ are chosen identical to the model $\widetilde{\Sigma}$, if follows that

$$\boldsymbol{x}_{\mathrm{F}}^*(0) = \boldsymbol{x}^*(0) \quad (5.37)$$

holds. With the IMC filter F implemented in an I/O normal form and with the purpose of the right inverse $\widetilde{\Sigma}$ to obtain the equality

$$\begin{bmatrix} \tilde{y}_{\mathrm{d}} \\ \vdots \\ \tilde{y}_{\mathrm{d}}^{(r-1)} \end{bmatrix} = \begin{bmatrix} x_{\mathrm{F},1}^{*} \\ \vdots \\ x_{\mathrm{F},r}^{*} \end{bmatrix} = \begin{bmatrix} x_{1}^{*} \\ \vdots \\ x_{r}^{*} \end{bmatrix} \tag{5.38}$$

(which is achieved using Eq. (4.22)), one finds that the first r states are equal due to the design of the right inverse and the last $n - r$ states are equal due to the request for identical internal dynamical behaviour (Lemma 5.1 holds true by definition in Corollary 5.3).

The result of Corollary 5.3 can now be used to define the requirements on the IMC filter F that is a feasible choice for NMP models $\widetilde{\Sigma}$ in such that it produces a stable I/O behaviour of $Q = \widetilde{\Sigma}^{\mathrm{r}} F$. Moreover, Corollary 5.3 can also be used to determine an internally stable implementation of the IMC controller Q. The following theorem summarises the preceding results and presents the main contribution of this section; that is, it states the requirements on F and presents the minimal realisation of the nonlinear IMC controller Q.

Theorem 5.3 (IMC design for non-minimum phase nonlinear systems using an exact model inverse). *Assume the model $\widetilde{\Sigma}$ and the filter F to be given as in Lemma 5.1. Additionally, input constraints $u \in \mathcal{U}$ of the model $\widetilde{\Sigma}$, as given in Eq. (5.1), exist.*
Assume the IMC filter F to be designed to have the following properties:

- *F is stable (in the sense that all trajectories of the states $\boldsymbol{x}_{\mathrm{F}}^*$ of F are non-explosive).*
- *F has identical trajectories of the internal dynamics as the model $\widetilde{\Sigma}$ in the sense of Lemma 5.1 (i. e., under a feedforward connection using the input u from Eq. (5.34)).*
- *The steady-state gain of F is one (i. e., $\lim_{t\to\infty} F\tilde{w}_{\mathrm{ss}} = \lim_{t\to\infty} \tilde{y}_{\mathrm{d}}(t) = \tilde{w}_{\mathrm{ss}}$ for $t \to \infty$).*

Then, an internally stable IMC controller Q for the model $\widetilde{\Sigma}$ is obtained by the IMC filter F (implemented in the I/O normal form) and the (algebraic) input calculation

$$u = \gamma\left(\boldsymbol{x}_{\mathrm{F}}^*, \tilde{y}_{\mathrm{d}}^{(r)}\right), \tag{5.39}$$

with $\tilde{y}_{\mathrm{d}}^{(r)}$ obtained from (cf. Eq. (4.19))

$$\tilde{y}_{\mathrm{d}}^{(r)} = \dot{x}_{\mathrm{F,r}} = a_{\mathrm{F}}\left(\boldsymbol{x}_{\mathrm{F}}^*\right) + b_{\mathrm{F}}\left(\boldsymbol{x}_{\mathrm{F}}^*\right)\tilde{w} \tag{5.40}$$

but limited by the saturation proposed in Theorem 5.1 with $\boldsymbol{x} = \boldsymbol{x}_{\mathrm{F}}^*$. With this IMC controller Q it follows that the output $\tilde{y}_{\mathrm{d}}$ of the IMC filter F is identical to the output $\tilde{y} = x_1^*$ of the model $\widetilde{\Sigma}$ ($\tilde{y}_{\mathrm{d}} = \tilde{y}$).

Proof. First, stability of the composition $Q = \widetilde{\Sigma}^{\mathrm{r}} F$ (i. e., stability of the I/O behaviour of the IMC controller Q) is shown:

By definition of F, the behaviour of its internal dynamics is identical to the behaviour of the internal dynamics of the model $\widetilde{\Sigma}$. Since a feedforward control structure depending on the model inverse is discussed, Corollary 5.3 is applicable, and it follows that the state trajectories $\boldsymbol{x}_{\mathrm{F}}^*$ of the IMC filter F are identical to the state trajectories $\boldsymbol{x}^*$ of the model $\widetilde{\Sigma}$ (i. e., $\boldsymbol{x}_{\mathrm{F}}^* = \boldsymbol{x}^*$). Due to the stability definition of F, it follows that $\boldsymbol{x}_{\mathrm{F}}^*$, and therewith $\boldsymbol{x}^*$, is non-explosive. Consequentially, $\tilde{y} = \tilde{y}_{\mathrm{d}}$ holds and is also non-explosive since h is analytic.

Second, stability of the implementation of the individual operators $\widetilde{\Sigma}^{\mathrm{r}}$ and F is shown (i. e., internal stability): Equation (5.39) differs from Eq. (5.34) in such that the states of the model inverse $\boldsymbol{x}$ have been replaced by the states $\boldsymbol{x}_{\mathrm{F}}^*$ of the IMC filter F. This is a feasible step, since the trajectories of those states are equivalent (i. e., $\boldsymbol{x}_{\mathrm{F}}^* = \boldsymbol{x}^*$). Since the trajectory $\boldsymbol{x}_{\mathrm{F}}$ is non-explosive, and since singularities are avoided by respecting the input constraints, the input u obtained by the algebraic equation (5.39) is finite (i. e., $u \in \mathcal{L}_\infty$). □

Theorem 5.3 implies an implementation of the IMC filter F which differs from the one proposed for minimum phase models (cf. Section 3.5 and Fig. 3.6). Here, the IMC filter F delivers all signals to the right inverse $\widetilde{\Sigma}$, defined by Eq. (5.39). Figure 5.9 displays these needed connection signals.

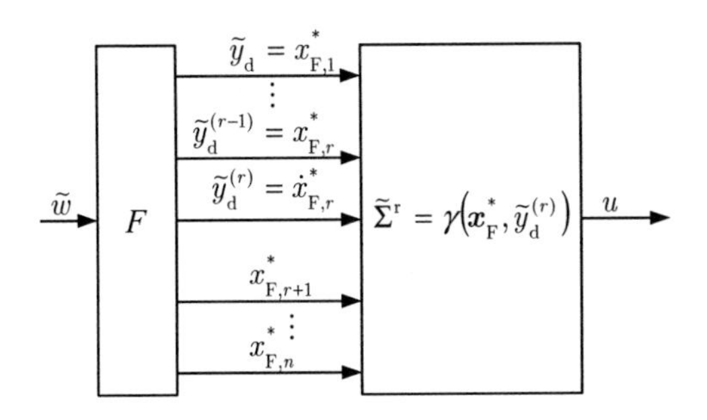

Fig. 5.9: Implementation signals of IMC filter F and right inverse $\widetilde{\Sigma}^{\mathrm{r}}$ in case of a non-minimum phase model.

Interpretation. Theorem 5.3 summarises the results of this section in such that it states the demands on the IMC filter F as well as the internally stable implementation of Q for IMC control of NMP systems.

The application of Theorem 5.3 is given in an algorithm below for reason of comparison to the IMC design for minimum phase systems as described in Algorithm 3.1:

Algorithm 5.1. ***NMP IMC design using a perfect inverse.***

Given: A stable and invertible NMP plant model $\widetilde{\Sigma}$.

Step 1: Compute the model inverse $\widetilde{\Sigma}^{\mathrm{r}}$ (specifically, find the necessary relationship corresponding to Eq. (5.34)).

Step 2: Find an IMC filter $F : \widetilde{\mathcal{W}} \to \widetilde{\mathcal{Y}}$ which

- *is stable,*
- *has a "good" step response from $\tilde{w}$ to $\tilde{y}$ with a steady-state gain of one (i. e., $\tilde{w}(t) = \tilde{y}_{\mathrm{d}}(t)$ for $t \to \infty$), and*
- *has identical internal dynamics as the model $\widetilde{\Sigma}$ in the sense of Lemma 5.1.*

Step 3: The IMC controller Q can now be implemented by F and Eq. (5.34)). The IMC control loop is given in Fig. 3.4 or, alternatively, by using Algorithm 3.2, also by substituting the internal model as in Fig. 3.8.

Result: Nonlinear output feedback IMC control loop.

The main difference between the generalised IMC design procedure of Algorithm 3.1 and the IMC design for NMP models in Algorithm 5.1 is the additional requirement on the IMC filter F to exhibit the identical internal dynamics as the model $\widetilde{\Sigma}$. The following example clarifies this concept and uses the results obtained above.

Example 5.6 (IMC of a nonlinear non-minimum phase plant):
Consider the plant model $\widetilde{\Sigma}$ with two states ($n = 2$) and a relative degree of $r = 1$

$$\widetilde{\Sigma}: \quad \dot{\boldsymbol{x}}^* = \begin{bmatrix} \frac{x_2^* + z\left(-(x_1^*)^2 u + (x_1^*)^3 + x_2^*(z+1)\right)}{z^2} \\ x_1^* u - \frac{x_2^* + z\left((x_1^*)^3 + x_2^*\right)}{z} \end{bmatrix}, \ \boldsymbol{x}^*(0) = \boldsymbol{x}_0^* \qquad (5.41)$$
$$\tilde{y} = x_1^*$$

in I/O normal form, with some parameter $z \in \mathbb{R}$ and $z \neq 0$. Hence, x_2^* is the (single) state of the internal dynamics. Computing the right inverse $\widetilde{\Sigma}^{\mathrm{r}}$ of the above model means solving the right-hand side of $\dot{x}_1^*$ with $\dot{x}_1^* \mapsto \dot{\tilde{y}}_{\mathrm{d}}$ for u, which yields

$$u = \frac{-z^2\dot{\tilde{y}}_{\mathrm{d}} + z(x_1^*)^3 + x_2^*(1+z) + z^2 x_2^*}{z(x_1^*)^2} \triangleq \gamma\left(\boldsymbol{x}^*, \tilde{y}_{\mathrm{d}}^{(r)}\right). \qquad (5.42)$$

Inserting Eq. (5.42) into Eq. (5.41) one finds the composition of right inverse $\widetilde{\Sigma}^{\mathrm{r}}$ with the plant model $\widetilde{\Sigma}$

$$\widetilde{\Sigma} \circ \widetilde{\Sigma}^{\mathrm{r}}: \quad \begin{aligned} \dot{\boldsymbol{x}}^* &= \begin{bmatrix} \dot{\tilde{y}}_{\mathrm{d}} \\ z x_2^* - z \dot{\tilde{y}}_{\mathrm{d}} \end{bmatrix}, \; \boldsymbol{x}^*(0) = \boldsymbol{x}_0^* \\ \tilde{y} &= x_1^*. \end{aligned} \tag{5.43}$$

The composition has, by design, linear dynamics of its first state x_1^* and, by chance, also a linear behaviour of the internal dynamics x_2^*. One sees that the pole of the internal dynamics is at z. Thus, for $z < 0$ the model $\widetilde{\Sigma}$ is minimum phase and for $z > 0$ it is NMP. Since it is desired to treat an NMP system, the parameter z is chosen as $z > 0$.

Consider an IMC filter F with input $\tilde{w}$

$$F: \quad \begin{aligned} \dot{\boldsymbol{x}}_{\mathrm{F}}^* &= \begin{bmatrix} \frac{x_{\mathrm{F},2}^*\left(z^2+2z\lambda+\lambda^2\right)}{z^2} + \frac{\lambda^2\left(-\tilde{w}+x_{\mathrm{F},1}^*\right)}{z} \\ -\frac{x_{\mathrm{F},2}^*\lambda(2z+\lambda)}{z} + \left(\tilde{w} - x_{\mathrm{F},1}^*\right)\lambda^2 \end{bmatrix}, \; \boldsymbol{x}_{\mathrm{F}}^*(0) = \boldsymbol{x}_0^* \\ \tilde{y}_{\mathrm{d}} &= x_{\mathrm{F},1}^* \\ \dot{\tilde{y}}_{\mathrm{d}} &= \dot{x}_{\mathrm{F},1}^*. \end{aligned} \tag{5.44}$$

This IMC filter F is stable and has a steady-state gain of one.

Using Lemma 5.1, it is shown that the behaviour of the internal dynamics of the IMC filter F is identical to $\widetilde{\Sigma}^{\mathrm{r}}$. This is done by subtracting the right-hand sides of $\dot{x}_{\mathrm{F},2}^*$ (cf. Eq. (5.44)) and $\dot{x}_2^*$ under inversion (cf. Eq. (5.43)) and the definition of $\dot{\tilde{y}}_{\mathrm{d}}$ given in Eq. (5.44). One finds that

$$\begin{aligned} &\overbrace{z\left(x_2^* - x_{\mathrm{F},2}^*\right) - 2x_{\mathrm{F},2}^*\lambda + \frac{1}{z}\left(z\tilde{w} - z x_{\mathrm{F},1}^* - x_{\mathrm{F},2}^*\right)\lambda^2 -}^{=\dot{x}_2^*} \\ &\underbrace{-2x_{\mathrm{F},2}^*\lambda + \frac{1}{z}\left(z\tilde{w} - z x_{\mathrm{F},1}^* - x_{\mathrm{F},2}^*\right)\lambda^2}_{=\dot{x}_{\mathrm{F},2}^*} = \underbrace{z\left(x_2^* - x_{\mathrm{F},2}^*\right)}_{\triangleq \delta(\boldsymbol{x}^*, \boldsymbol{x}_{\mathrm{F}}^*)} \end{aligned} \tag{5.45}$$

holds. It follows from Lemma 5.1 that the internal dynamics of the filter F and the right inverse $\widetilde{\Sigma}^{\mathrm{r}}$ are identical and, consequentially, Corollary 5.3 states that all states of F and $\widetilde{\Sigma}^{\mathrm{r}}$ are identical.

Finally, according to Theorem 5.3, the desired IMC controller Q is defined by the filter F given by Eq. (5.44) and the input calculation from Eq. (5.42) but using the states of the filter F, namely

$$\begin{aligned} u &= \gamma\left(\boldsymbol{x}^*, \dot{\tilde{y}}_{\mathrm{d}}\right) \text{ with } \boldsymbol{x}^* \mapsto \boldsymbol{x}_{\mathrm{F}}^*, \; \dot{\tilde{y}}_{\mathrm{d}} \mapsto \dot{x}_{\mathrm{F},1}^* \\ &= \frac{z \cdot (x_{\mathrm{F},1}^*)^3 - z^2\tilde{w}\lambda^2 + z^2 x_{\mathrm{F},1}^*\lambda^2 + x_{\mathrm{F},2}^*\left(1 + z + z^2 + 2z^2\lambda + z\lambda^2\right)}{z \cdot (x_{\mathrm{F},1}^*)^2} \end{aligned} \tag{5.46}$$

Thus, the IMC controller Q can be interpreted to be the filter F with the new output u given by Eq. (5.46).

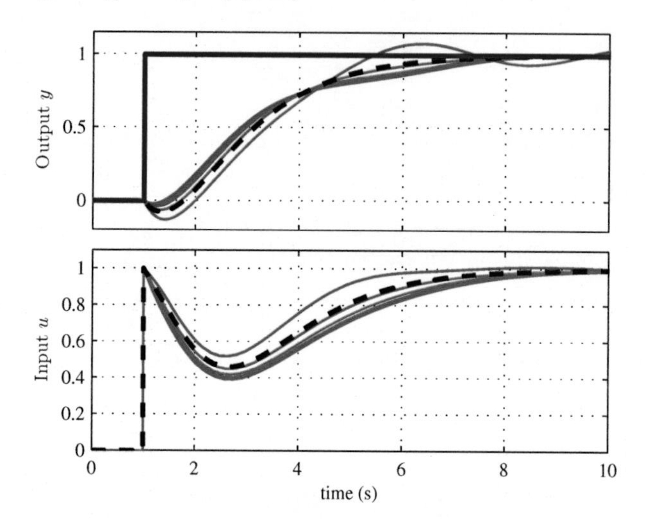

Fig. 5.10: Closed-loop simulation with a reference step occurring at time $t = 1\text{s}$. Nominal values are $\lambda = 1$ and $z = 2$. The nominal response is plotted in the bold dashed line. Responses with only the plant parameter z adjusted between 1.3 and 4 are plotted in thin solid lines.

Figure 5.10 shows the simulation results from a reference step occurring at $t = 1\text{s}$. The filter pole is chosen as $\lambda = 1$ and in the nominal case z is at $z = 2$.

The following can be seen from Fig. 5.10: The bold dashed line indicates the nominal plant output y and nominal plant input (IMC controller output) u. In the nominal case, the output $\tilde{y}_{\mathrm{d}}$ of the IMC filter F and the output y of the plant are equal ($\tilde{y}_{\mathrm{d}} = y$). The filter output shows an NMP characteristic (inverse response) due to the zero at z. As expected, the input trajectory u is non-explosive despite the perfect inversion of the NMP nonlinear model $\widetilde{\Sigma}$.

In order to explicitly show that this approach does not rely on a perfect inversion (no pole-zero cancellation) of the plant Σ, the plant's NMP behaviour is altered, maintaining the model $\widetilde{\Sigma}$ and, therewith, nominal IMC controller Q. This is done by changing z in the plant equations between 1.4 and 5 in increments of 0.8. Thus, there is a significant modelling error concerning the NMP characteristic. Despite this, the closed-loop behaviour of the IMC is good and, as expected, this is achieved with a non-explosive input signal u in all cases. ■

5.3.4 Construction of the Non-Minimum Phase IMC Filter

Despite the successful application of Theorem 5.3 in the example above, there is still one significant open problem: How this IMC filter F is obtained. This section addresses this problem by showing that F can be obtained by the solution of a control problem.

The following constructs the structure of F and defines the control problem. This is done in three steps:

Algorithm 5.2 (Construction of an NMP filter F).

Given: An NMP nonlinear model $\widetilde{\Sigma}$.

Step 1 (Structure of F): Assume the model $\widetilde{\Sigma}$ to be given in I/O normal form (4.19)*, shown in Fig. 5.7.*

Construct the IMC filter F with the same number of states n, the same relative degree r in the same I/O normal form (4.19) *but with all variables defined with a subscript* F*, the output $\tilde{y} = x^*_{\mathrm{F},1}$ and the input $\tilde{w}$.*

Step 2 (Internal Dynamics of F): Insert the inverting input (5.34) *into the internal dynamics of the plant model $\widetilde{\Sigma}$. This results in*

$$[\dot{x}^*_{r+1}, \cdots, \dot{x}^*_n]^T = \boldsymbol{p}(\boldsymbol{x}^*) + \boldsymbol{q}(\boldsymbol{x}^*) \cdot \gamma\left(\boldsymbol{x}^*, \tilde{y}_{\mathrm{d}}^{(r)}\right). \qquad (5.47)$$

Since, according to Corollary 5.3, the states $\boldsymbol{x}^*_{\mathrm{F}}$ of the filter F and the states of the inverted model $\widetilde{\Sigma}^{\mathrm{r}}$ are going to be identical, one finds the states of the internal dynamics of the IMC filter F from Eq. (5.47) *with the substitution*

$$\boldsymbol{x}^* \mapsto \boldsymbol{x}^*_{\mathrm{F}}, \text{ the input } (5.34) \text{ and}$$
$$\tilde{y}_{\mathrm{d}}^{(r)} = \dot{x}^*_{\mathrm{F},r} = a_{\mathrm{F}}(\boldsymbol{x}^*_{\mathrm{F}}) + b_{\mathrm{F}}(\boldsymbol{x}^*_{\mathrm{F}})\tilde{w}$$

as

$$\begin{bmatrix} \dot{x}^*_{\mathrm{F},r+1} \\ \vdots \\ \dot{x}^*_{\mathrm{F},n} \end{bmatrix} = \boldsymbol{p}(\boldsymbol{x}^*_{\mathrm{F}}) + \boldsymbol{q}(\boldsymbol{x}^*_{\mathrm{F}}) \cdot \frac{-a(\boldsymbol{x}^*_{\mathrm{F}}) + a_{\mathrm{F}}(\boldsymbol{x}^*_{\mathrm{F}}) + b_{\mathrm{F}}(\boldsymbol{x}^*_{\mathrm{F}})\tilde{w}}{b(\boldsymbol{x}^*_{\mathrm{F}})}. \qquad (5.48a)$$

Step 3 (F as the solution of a control problem): The remaining states of F are (cf. Eq. (4.19))

$$\dot{x}^*_{\mathrm{F},i} = x^*_{\mathrm{F},i+1} \quad \textit{for} \ \ i = 1, \ldots, r-1 \tag{5.48b}$$

$$\dot{x}^*_{\mathrm{F},r} = a_{\mathrm{F}}(\boldsymbol{x}^*_{\mathrm{F}}) + b_{\mathrm{F}}(\boldsymbol{x}^*_{\mathrm{F}})\tilde{w}. \tag{5.48c}$$

The solution of the following control problem gives the desired IMC filter F with NMP behaviour:

Control Problem (**Construction of an NMP IMC filter** F). The system to be controlled (i. e., the *plant* in this context) consists of the integrator chain (5.48b) and the internal dynamics (5.48a). The *state feedback controller* to this plant is given by (5.48c).

The plant is perfectly known and no disturbances need to be considered since the complete IMC filter is an artificial operator. The closed-loop behaviour of the plant (5.48a)-(5.48b) and the controller (5.48c) is the IMC filter F.

The control problem requires to *find some functions* $a_{\mathrm{F}}(\boldsymbol{x}^*_{\mathrm{F}})$ and $b_{\mathrm{F}}(\boldsymbol{x}^*_{\mathrm{F}})$ such that

- the closed-loop system is stable, and
- that the steady-state gain of F is one (i. e., $\tilde{w}(t) = \tilde{y}_{\mathrm{d}}(t)$ for $t \to \infty$).

Result: The result is an IMC filter F that meets all requirements of Theorem 5.3.

In this thesis, the control problem given above is not solved for the general case. This is left as an open problem to be solved for each application individually.

5.4 Treating Measured Disturbances

In some cases, a disturbance is measured and, thus, is known. Such disturbances are denoted by d_{m}. The effect of a measured disturbance can be attenuated by the IMC controller Q, directly (i. e., the attenuation is not based on the feedback signal but can be initiated in the feedforward path), since the disturbance value is available. In some cases, its effect can be removed completely in the feedforward path. The following is based on [42].

Main idea. It is proposed to model the effect of the measured disturbance d_{m}:

$$\widetilde{\Sigma}: \dot{\boldsymbol{x}}(t) = \boldsymbol{f}\left(\boldsymbol{x}(t), d_{\mathrm{m}}(t), u(t)\right), \quad \boldsymbol{x}(0) = \boldsymbol{x}_0,\ \boldsymbol{x} \in \mathcal{X}, \tag{5.49a}$$

$$\tilde{y}(t) = h\left(\boldsymbol{x}(t), d_{\mathrm{m}}(t), u(t)\right), \quad u \in \mathcal{U},\ \tilde{y} \in \widetilde{\mathcal{Y}}. \tag{5.49b}$$

The operator of the plant model $\widetilde{\Sigma}$ can now be written as

$$\widetilde{\Sigma}: \quad \mathcal{U} \times \mathcal{D}_{\mathrm{m}} \to \widetilde{\mathcal{Y}}, \tag{5.50}$$

and maps the input trajectory $u \in \mathcal{U}$ *and* the trajectory of the measured disturbance $d_{\mathrm{m}} \in \mathcal{D}_{\mathrm{m}}$ into the output trajectory $\tilde{y} \in \widetilde{\mathcal{Y}}$. The right inverse $\widetilde{\Sigma}^{\mathrm{r}}$ of this operator is defined along the lines of Definition 3.4 and with the purpose of the control design in mind as

$$\widetilde{\Sigma}^{\mathrm{r}}: \quad \widetilde{\mathcal{Y}} \times \mathcal{D}_{\mathrm{m}} \to \mathcal{U} \tag{5.51a}$$

and fulfils

$$\widetilde{\Sigma} \circ \left(\widetilde{\Sigma}^{\mathrm{r}} \circ \left(\tilde{y}_{\mathrm{d}}, d_{\mathrm{m}}\right), d_{\mathrm{m}}\right) = \tilde{y}_{\mathrm{d}}. \tag{5.51b}$$

Thus, the plant model $\widetilde{\Sigma}$ and its right inverse $\widetilde{\Sigma}^{\mathrm{r}}$ are defined equivalently to Definition 3.4 on page 50 with the difference being that the information of the disturbance d_{m} enters both. According to the Definition of the right inverse above, the effect of the disturbance will be cancelled completely. Figure 5.11 shows such a feedforward control structure.

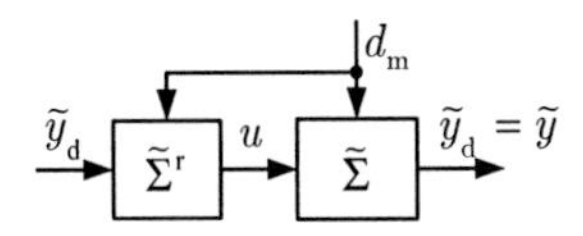

Fig. 5.11: Right inverse with measured disturbances.

Design. If one extends the notation in the previous section from $\boldsymbol{f}(\boldsymbol{x}, u)$ into $\boldsymbol{f}(\boldsymbol{x}, d_{\mathrm{m}}, u)$ then the resulting right inverse will automatically be obtained with the methods reviewed in Chapter 4. Hence, the effect of measured disturbance will automatically be cancelled by the right inverses of flat and input-affine systems if the dependency of $\boldsymbol{f}(\boldsymbol{x}, d_{\mathrm{m}}, u)$

and $h(\boldsymbol{x}, d_{\mathrm{m}}, u)$ on the measured disturbance d_{m} is considered appropriately. For this reason, the Lie derivative needs to be altered to

$$\begin{aligned} L_f^k h(\boldsymbol{x}, d_{\mathrm{m}}, u) &= \frac{\partial}{\partial \boldsymbol{x}} \left\{ L_f^{k-1} h(\boldsymbol{x}, d_{\mathrm{m}}, u) \right\} \boldsymbol{f}(\boldsymbol{x}, d_{\mathrm{m}}, u) \\ L_f^0 h(\boldsymbol{x}, d_{\mathrm{m}}, u) &= h(\boldsymbol{x}, d_{\mathrm{m}}, u). \end{aligned} \tag{5.52}$$

The resulting IMC structure, which respects a measured disturbance, is shown in Fig. 5.12.

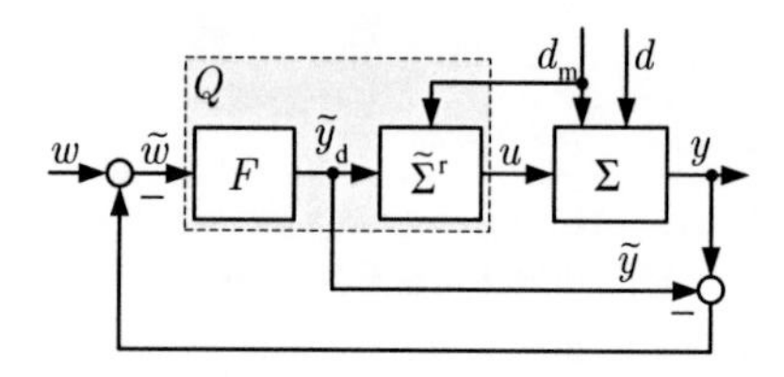

Fig. 5.12: IMC structure with a measured disturbance d_{m} and unmeasured disturbances d.

However, depending on where the disturbance enters the model, it cannot always be cancelled if it is only known up to the current time t. To clarify this statement, the following introduces a relative degree ρ of the measured disturbance d_{m}:

$$\rho = \min_k \left\{ \frac{\partial}{\partial d_{\mathrm{m}}} L_f^k h(\boldsymbol{x}, d_{\mathrm{m}}, u) \neq 0 \right\} \tag{5.53}$$

With the relative degree r of the input u one find the following two cases (cf. [42]):

$r \leq \rho$: If the relative degree r of the input u is smaller or equal to the relative degree ρ of the measured disturbance d_{m} then the right inverse $\widetilde{\Sigma}^{\mathrm{r}}$ will depend on the current measured disturbance $d_{\mathrm{m}}(t)$. Additionally, $\widetilde{\Sigma}^{\mathrm{r}}$ will *not* depend on differentiations of $d_{\mathrm{m}}(t)$, like $\dot{d}_{\mathrm{m}}(t)$.

In other words: The input u affects the output $\tilde{y}$ more directly than the measured disturbance d_{m}.

$r > \rho$: If the relative degree r of the input u is greater than the relative degree ρ of the measured disturbance d_{m} then the right inverse $\widetilde{\Sigma}^{\mathrm{r}}$ will depend on $r - \rho$ differentiations of the measured disturbance. Hence it will need the signal $d_{\mathrm{m}}^{(r-\rho)}(t)$ and its lower derivatives.

In other words: The measured disturbance d_{m} affects the output $\tilde{y}$ more directly than the input u. Hence, the disturbance cannot be cancelled unless its derivatives are known.

In most control problems, derivatives of a measured disturbance are not available. Thus, in the case $r > \rho$, the measured disturbance cannot be cancelled perfectly since the necessary signals (namely $d_{\mathrm{m}}^{(r-\rho)}(t)$ and its lower derivatives) are not available. In such a case, however, it is proposed to generate an approximation of $d_{\mathrm{m}}^{(r-\rho)}(t)$ and its lower derivatives by feeding d_{m} into an SVF of relative order $r-\rho$. An SVF is especially useful since the designer can choose the bandwidth of the SVF as a compromise between noise amplification and phase lag.

Fortunately, the control problems discussed in Chapters 6 and 7 both belong to the case $r > \rho$, where no differentiations of the measured disturbances are necessary.

5.5 IMC of Simple Quadratic Nonlinear MIMO Systems

It is not desired to treat MIMO systems extensively. The following merely gives a brief discussion on some structural changes in the IMC design for simple quadratic systems with p inputs and outputs, without claiming to be an exhaustive examination. For an in-depth treatment on MIMO inverses, the reader is referred to [20, 48, 51, 80, 81] and the references therein.

First, some nomenclature for MIMO systems is given. Consider the MIMO model (3.2) with p inputs $\boldsymbol{u} = [u_1, \ldots, u_p]^T$ and p outputs $\tilde{\boldsymbol{y}} = [\tilde{y}_1, \ldots, \tilde{y}_p]^T$. The relative degree $\boldsymbol{r}$ is a p-vector. Loosely speaking, the value r_j says that the r_j-th derivative $\tilde{y}_k^{(r_j)}$ of the k-th output $\tilde{y}_k$ depends on some input u_j (with $j \in \{1, 2, \ldots, m\}$) which no other output derivative uses to define its relative degree. Such a relationship is to be established (if possible) for each entry in the $\boldsymbol{u}$ vector for a different output.

Pre-integration. One finds that, in general, the right inverse $\widetilde{\Sigma}^{\mathrm{r}}$ of a MIMO nonlinear model depends not only on the demanded trajectory $\tilde{\boldsymbol{y}}_{\mathrm{d}}$ and some of its derivatives, but also on the input vector $\boldsymbol{u}$ *and* some of its derivatives $\dot{\boldsymbol{u}}, \ddot{\boldsymbol{u}}, \ldots$. This is established in e. g., [80] for general nonlinear systems, can be seen from the I/O normal form of input-affine systems (see e. g., [20, 48, 51]), and the definition of differential flatness of MIMO systems (see e.g., [81]). A typical solution to generate derivatives for some inputs is pre-integration [80]. Suppose l derivatives of u_j are needed. Then choose ${u_j}^{(l)}$ as a new input ν_j and extend the state-space

$$\begin{aligned}\dot{x}_{n+1} &= x_{n+2} = u_j \\ &\vdots \\ \dot{x}_{n+l-1} &= x_{n+l} = {u_j}^{(l-1)} \\ \dot{x}_{n+l} &= {u_j}^{(l)} = \nu.\end{aligned}$$

The relative degree for input u_j to output $\tilde{y}_k$ changes from r_j to $r_j + l$, with respect to the new input ν_j. In the following, it is assumed that such a pre-integration has been performed and the new input and the new relative degree are, again, denoted by $\boldsymbol{u}$ or $\boldsymbol{r}$, respectively. Thus, input derivatives are not considered any further.

IMC filter. An inversion of a MIMO system $\widetilde{\Sigma}$ results in an I/O decoupling. This is obvious from the composition $\widetilde{\Sigma} \circ \widetilde{\Sigma}^{\mathrm{r}} = \underset{\sim}{I}$, which is implied by the definition (Definition 3.4) of the right inverse $\widetilde{\Sigma}^{\mathrm{r}}$. Therefore, the MIMO IMC filter $\boldsymbol{F}$ needs to be a diagonal $p \times p$ operator

$$\boldsymbol{F} = \begin{bmatrix} F_1 & 0 & 0 \\ 0 & \ddots & 0 \\ 0 & 0 & F_p \end{bmatrix}, \tag{5.54}$$

which maps the vector signal $\tilde{\boldsymbol{w}}$ into the signal $\tilde{\boldsymbol{y}}_{\mathrm{d}}$. Each entry $F_1, \cdots, F_p$ is designed and implemented as proposed in Section 3.5.1 for the SISO case.

In conclusion, the MIMO case does not change the philosophy of how an IMC controller is designed. The only changes lie in the inversion which might have to be extended by the concept of pre-integration and that $\boldsymbol{F}$ is a MIMO diagonal operator.

Input constraints. Input constraints in the SISO case are considered (cf. Section 5.1) by establishing a relationship between the input u and the highest derivative $\tilde{y}_{\mathrm{d}}^{(r)}$ of the filter output $\tilde{y}_{\mathrm{d}}$. In the MIMO case, this relationship contains more entries of the input vector $\boldsymbol{u}$. Assuming that the necessary pre-integration has been performed, then this relationship can, in general, be given as

$$\tilde{y}_{\mathrm{d},k}^{(r_j)} = \varphi_j(\boldsymbol{x}, u_1, \ldots, u_k, \ldots, u_m) \text{ for } k = 1, \ldots, p. \tag{5.55}$$

Thus, if not all desired highest derivatives $\tilde{y}_{\mathrm{d},k}^{(r_j)}$ of the output vector $\tilde{\boldsymbol{y}}_{\mathrm{d}}$ of the filter $\boldsymbol{F}$ can be reached with permissible inputs $\boldsymbol{u}_{\min} \leq \boldsymbol{u} \leq \boldsymbol{u}_{\max}$ then, in general, there is no unique solution as to which derivative (or combination of) $\tilde{y}_{\mathrm{d},k}^{(r_j)}$ to limit. Thus, in this case, the engineer is left with an infinite number of possibilities as to which derivatives to limit. Currently, there is no generally applicable solution to this problem.

Let $\boldsymbol{\varphi} = [\varphi_1, \ldots, \varphi_p]^T$. Note that a unique solution exists if the condition

$$\mathrm{Rank}\frac{\partial \boldsymbol{\varphi}(\boldsymbol{x}, \boldsymbol{u})}{\partial \boldsymbol{u}} = p \tag{5.56}$$

holds. This is the case if, for example, the $\varphi_j(\cdot)$ can be sorted such that a lower or upper triangular form of the inputs $u_1, \ldots, u_m$ can be achieved. For example, the following relationships

$$\begin{aligned}
\tilde{y}_{\mathrm{d},1}^{(r_1)} &= \varphi_1(\boldsymbol{x}, u_1) \\
\tilde{y}_{\mathrm{d},2}^{(r_2)} &= \varphi_2(\boldsymbol{x}, u_1, u_2) \\
\tilde{y}_{\mathrm{d},3}^{(r_3)} &= \varphi_3(\boldsymbol{x}, u_1, u_2, u_3)
\end{aligned}$$

present a lower triangular form in the inputs u_1, u_2, u_3 and, thus, a limitation in $\boldsymbol{u}$ can be mapped to a unique solution for a limitation in $\tilde{\boldsymbol{y}}_{\mathrm{d}}$.

5.6 Summary

The basic design procedure of IMC, as introduced in Section 3.5, has a rather limited system class, namely minimum phase models with well-defined relative degree and arbitrary but finite input signals. This chapter extended the system class to

- non-minimum phase models,
- models with an ill-defined relative degree,
- models with input constraints, and
- models with measured disturbances.

Most extensions exploit Proposition 3.2 . Thus, the IMC design for this extended system class fits perfectly into a larger picture and can, therefore, be regarded as a generalisation of the basic IMC design procedure as introduced in Section 3.5.

Part II

INTERNAL MODEL CONTROL OF TURBOCHARGED ENGINES

This part presents the practical contribution of this thesis. It applies the proposed nonlinear internal model control to two automotive control problems. Those are the boost pressure control of a one-stage turbocharged diesel engine and the control of three pressures of a two-staged turbocharged diesel engine. The proposed controller for a two-stage turbocharged diesel engine is the first solution of its kind.

6. CONTROL OF A ONE-STAGE TURBOCHARGED DIESEL ENGINE

This chapter develops a SISO IMC boost pressure controller for a turbocharged diesel engine with a variable-nozzle turbine (VNT) (see Fig. 6.1). For this purpose, the basic IMC design procedure (cf. Section 3.5) is employed where the model inverse is computed using the flat model $\widetilde{\Sigma}$ (cf. Section 4.1). The design is extended to handle limited constraints (cf. Section 5.1) and treats measured disturbances (cf. Section 5.4).

6.1 *Function of a One-Stage Turbocharged Diesel Engine*

The function of an internal combustion engine is described in detail in e. g., [47]. The article [25] focuses on the specifics of diesel engines. The various methods of forced air induction are described in [105] and [27]. The work of [26] focuses on turbocharging itself.

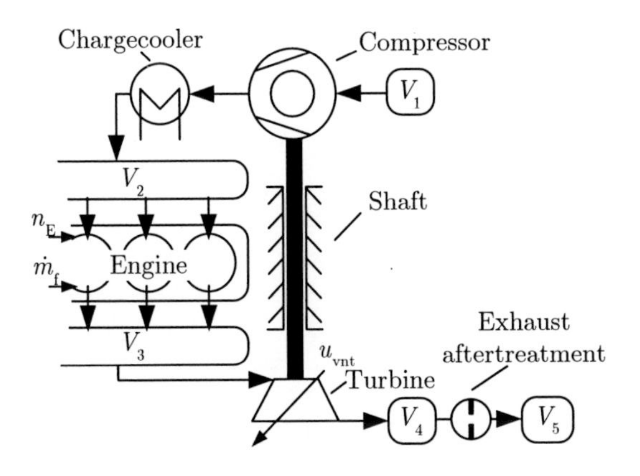

Fig. 6.1: Air-system of a one-stage turbocharged engine.

Figure 6.1 shows the air-system of a common on-stage turbocharged engine, using a VNT [26]. The engine speed is denoted by n_{E} and the injected fuel mass flow is denoted by $\dot{m}_{\mathrm{f}}$.

In the following, the air path through the system is described, starting with air from the environment V_1. In the one-stage turbocharged engine with a VNT, fresh air is aspirated from the environment V_1 and is compressed by the compressor. Before the compressed air enters the intake manifold V_2, it is cooled down by the intercooler that simultaneously increases the air density. The pressure in the intake manifold V_2 is called the "boost pressure" p_2. The boost pressure p_2 forces the air into the engine where it is mixed with fuel and, in turn, leads to combustion. The hot exhaust gas is pushed into the exhaust manifold V_3. Its pressure is called the "exhaust back pressure" p_3. The exhaust gas is led over the turbine, whose nozzle geometry u_{vnt} affects the amount of exhaust energy that is converted to drive the compressor over the shaft. Finally, the exhaust gas flows through the exhaust pipe V_4, through the aftertreatment (e. g., catalytic converters), and back into the environment V_5. The engine speed n_{E} and the injected fuel-mass flow $\dot{m}_{\mathrm{f}}$ deliver energy to the air-system. They can be considered to be exogenous variables, stimuli, or disturbances, depending on the problem to be solved (e. g., parameter identification or control design).

This method of forced induction is the most widely used for diesel engines. Due to their higher combustion temperatures, spark ignition engines pose higher demands on materials used for the variable nozzles. Currently, only the Porsche 911 Turbo (type 997) [7] uses this technology. Other turbocharged spark ignition engines use a wastegate turbine instead of the variable nozzle turbine.

6.2 Control Problem

The aim of turbocharging is to induce a desired amount of oxygen into the combustion chamber. Since this value is correlated to the measured boost pressure p_2, the boost pressure becomes the main control variable. The input is the nozzle geometry ($u = u_{\mathrm{vnt}}$). Engine speed n_{E} and injected fuel mass flow $\dot{m}_{\mathrm{f}}$ are assumed to be exogenous variables and are treated as measured disturbances ($\boldsymbol{d}_{\mathrm{m}} = [n_{\mathrm{E}},\ \dot{m}_{\mathrm{f}}]^T$). The behaviour of the one-stage air-system can be described using a model $\widetilde{\Sigma}$ which is given by Eq. (6.1). The input and output signals, including the measured disturbances $\boldsymbol{d}_{\mathrm{m}}$ are shown in Fig. 6.2.

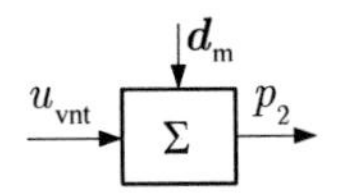

Fig. 6.2: Input and output signals of a one-stage turbocharged engine.

Control Problem (one-stage). The measured variable is the boost pressure $y = p_2$ and the control input is the VNT position $u = u_{\text{vnt}}$. The controller should ensure that

- the boost pressure p_2 tracks a constant reference signal w with zero steady-state offset,
- the input constraint $0 = u_{\min} \leq u_{\text{vnt}} \leq u_{\max} = 1$ of the VNT position is not violated,
- the effect of the measured disturbances $\boldsymbol{d}_{\text{m}}$ is compensated, and
- robust stability is guaranteed.

Unfortunately, the precise requirements for the behaviour of the controlled system cannot be reviewed here in detail for reasons of confidentiality. In general, performance requirements concern mainly a desired speed, a demand on overshoot of boost pressure where less overshoot is better, and a demand in undershoot of boost pressure where less undershoot is better. Typically, the permissible overshoot in the boost-pressure is requested to be at most 0.15 bar and the undershoot is requested to be less than 0.03 bar. Note, however, that even the series production controller violates these demands frequently, as will be shown later. Hence, those requirements should be interpreted more as a guideline than as as stringent demands.

Model. The derivation of the model equations is given in Appendix A. For the sake of better readability, the equations of the resulting control-oriented model are repeated here.

The input u is a function of the nozzle position of the variable nozzle turbine (VNT). The engine speed n_{E} and the injected fuel mass $\dot{m}_{\text{f}}$ are considered to be measured disturbances $\boldsymbol{d}_{\text{m}} = [n_{\text{E}}, \dot{m}_{\text{f}}]^T$. The model $\widetilde{\Sigma}$ is given by

$$\begin{aligned}\widetilde{\Sigma}: \dot{x}_1 =& \frac{k_1 k_2}{x_1}\left(k_3(\boldsymbol{d}_\mathrm{m}) + k_6(\boldsymbol{d}_\mathrm{m}) k_4 x_2\right)\varphi_1(x_2) - \frac{k_1 k_5 \varphi_3(x_2)}{x_1}\varphi_2(x_1, x_2)\\ &- \frac{k_1 k_2}{x_1}\left(k_3(\boldsymbol{d}_\mathrm{m}) + k_6(\boldsymbol{d}_\mathrm{m}) k_4 x_2\right)\varphi_1(x_2) u\\ \dot{x}_2 =& \frac{\varphi_4(x_2)}{k_7}\varphi_2(x_1, x_2) - \frac{k_9}{k_7} k_6 x_2\end{aligned} \tag{6.1a}$$

with the initial condition $\boldsymbol{x}(0) = \boldsymbol{x}_0$ and the output equation

$$\tilde{y} = x_2 = p_2 \tag{6.1b}$$

with

$$\begin{aligned}\varphi_1(x_2) &= \frac{k_9}{k_4} + \frac{k_{17}(\boldsymbol{d}_\mathrm{m})}{k_{14}(\boldsymbol{d}_\mathrm{m}) + k_4 x_2} \triangleq T_3\\ \varphi_2(x_1, x_2) &= k_8 \frac{k_{10} x_1^2 - \varphi_3(x_2)}{k_{11} x_1} \triangleq \dot{m}_\mathrm{C,out}\\ \varphi_3(x_2) &= k_{16}\left(\left(\frac{x_2}{p_\mathrm{amb}}\right)^{k_{12}} - 1\right) \triangleq \frac{\dot{H}_\mathrm{C,out} + \dot{H}_\mathrm{C,in}}{\dot{m}_\mathrm{C,out}}\\ \varphi_4(x_2) &= k_{15}\varphi_3(x_2) + k_{13} \triangleq T_2,\end{aligned}$$

where the coefficients k_i are system parameters (e. g., diameters, inertia of the turbocharger, etc.), p_amb is the ambient pressure and the $\varphi_j(\cdot)$ are some nonlinear functions of the states. The model is stable and minimum phase and, thus, can be used for IMC design.

6.3 Nonlinear IMC of the Air-System

6.3.1 Model Inverse

For the design model (6.1), it is shown that the boost pressure p_2 is a flat output as well as the system output y

$$z = y = p_2. \tag{6.2}$$

In order to show the flatness of the system, the state variables $\boldsymbol{x} = [\omega, p_2]^T$ as well as the input u have to be expressed as a function of z and a finite number of its time derivatives (cf. Eqns. (4.1)-(4.4)). Equation (6.2) relates the flat output z to the first state, namely the boost pressure p_2. By differentiating Eq. (6.2) along the model dynamics (6.1a), one gets

$$\begin{aligned} \dot{z} &= \frac{\varphi_4(x_2)}{k_7}\varphi_2(x_1, x_2) - \frac{k_9}{k_7}k_6(\boldsymbol{d}_\mathrm{m})x_2 \\ &= \frac{\varphi_4(z)k_8\left(k_{10}x_1^2 - \varphi_3(z)\right)}{k_7 k_{11} x_1} - \frac{k_9}{k_7}k_6(\boldsymbol{d}_\mathrm{m})z. \end{aligned} \tag{6.3}$$

The quadratic equation Eq. (6.3) has the following positive unique solution for the second state, the turbocharger speed ω, given that $z > p_\mathrm{amb}$, i. e. $p_2 > p_\mathrm{amb}$ holds:

$$\begin{aligned} x_1 =& \frac{k_7 k_{11}\dot{z} + \frac{k_9}{k_7}k_6(\boldsymbol{d}_\mathrm{m})z}{2k_8 k_{10}\varphi_4(z)} \\ &+ \frac{\sqrt{(k_7 k_{11}\dot{z} + \frac{k_9}{k_7}k_6(\boldsymbol{d}_\mathrm{m})z)^2 + 4k_8^2 k_{10}\varphi_3(z)\varphi_4(z)}}{2k_8 k_{10}\varphi_4(z)} \end{aligned} \tag{6.4}$$

Equation (6.4) expresses the turbocharger speed $\omega = x_1$ to the flat output z and its first derivative $\dot{z}$. Using the second derivative of Eq. (6.2) and taking Eq. (6.4) into account, one gets

$$\ddot{z} = \varphi_3(z, \dot{z}, \boldsymbol{d}_\mathrm{m}) + \varphi_5(z, \dot{z}, \boldsymbol{d}_\mathrm{m})u, \tag{6.5}$$

which leads to:

$$u = \underbrace{\frac{\ddot{z} - \varphi_3(z, \dot{z}, \boldsymbol{d}_\mathrm{m})}{\varphi_5(z, \dot{z}, \boldsymbol{d}_\mathrm{m})}}_{\psi_\mathrm{u}(z,\dot{z},\boldsymbol{d}_\mathrm{m})}; \quad \varphi_5(z, \dot{z}, \boldsymbol{d}_\mathrm{m}) \neq 0. \tag{6.6}$$

A solution exists for all z, $\dot{z}$, if $z \geq p_\mathrm{amb}$. This is always the case, since the lowest attainable boost pressure $p_2 = y = z$ is the ambient pressure p_amb. The right inverse $\widetilde{\Sigma}^\mathrm{r}$ of the model of a one-stage turbocharged air-system can be obtained from Eq. (6.6) by letting $z = y_\mathrm{d}$

$$\widetilde{\Sigma}^\mathrm{r}: \quad u = \frac{\ddot{\tilde{y}}_\mathrm{d} - \varphi_3(\tilde{y}_\mathrm{d}, \dot{\tilde{y}}_\mathrm{d}, \boldsymbol{d}_\mathrm{m})}{\varphi_5(\tilde{y}_\mathrm{d}, \dot{\tilde{y}}_\mathrm{d}, \boldsymbol{d}_\mathrm{m})} \tag{6.7}$$

Since the actual output y is equal to the flat output z (cf. Eq. (6.2)) no differential equation (i. e., $F_{\mathrm{y}\to\mathrm{z}} = I$) needs to be solved to calculate one from the other.

6.3.2 IMC Filter

The IMC filter needs to create the trajectory for the desired output $\tilde{y}_\mathrm{d}(t)$ and the first two derivatives $\dot{\tilde{y}}_\mathrm{d}(t)$ and $\ddot{\tilde{y}}_\mathrm{d}(t)$. The transfer function of the filter F from its input signal $\tilde{w}$ to the output trajectory y_d is chosen as proposed in Section 3.5 by

$$\frac{\tilde{y}_\mathrm{d}(s)}{\tilde{w}(s)} = \frac{1}{(s/\lambda + 1)^2}. \tag{6.8}$$

The input constraints

$$u_\mathrm{min} \leq u \leq u_\mathrm{max}. \tag{6.9}$$

have to be considered. This is done according to Section 5.1.

From Eq. (6.5) one finds maximum and minimum permissible second derivatives of $\tilde{y}_\mathrm{d}$

$$\begin{aligned} \ddot{\tilde{y}}_\mathrm{max} =& \varphi_3(\tilde{y}, \dot{\tilde{y}}, n, \dot{m}_\mathrm{f}) + \\ & \max(\varphi_5(\tilde{y}, \dot{\tilde{y}}, n, \dot{m}_\mathrm{f}) u_\mathrm{max}, \varphi_5(\tilde{y}, \dot{\tilde{y}}, n, \dot{m}_\mathrm{f}) u_\mathrm{min}) \\ \ddot{\tilde{y}}_\mathrm{min} =& \varphi_3(\tilde{y}, \dot{\tilde{y}}, n, \dot{m}_\mathrm{f}) + \\ & \min(\varphi_5(\tilde{y}, \dot{\tilde{y}}, n, \dot{m}_\mathrm{f}) u_\mathrm{max}, \varphi_5(\tilde{y}, \dot{\tilde{y}}, n, \dot{m}_\mathrm{f}) u_\mathrm{min}). \end{aligned} \tag{6.10}$$

The limited IMC filter is shown in Fig. 6.3.

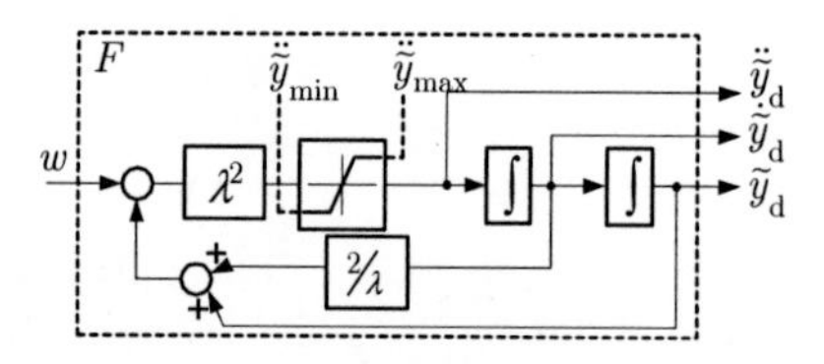

Fig. 6.3: IMC filter F with saturation to respect input constraints.

6.3.3 Complete IMC Law

The resulting IMC feedback structure is presented in Fig. 6.4 with F from Fig. 6.3 and $\tilde{\Sigma}^\mathrm{r}$ from Eq. (6.7).

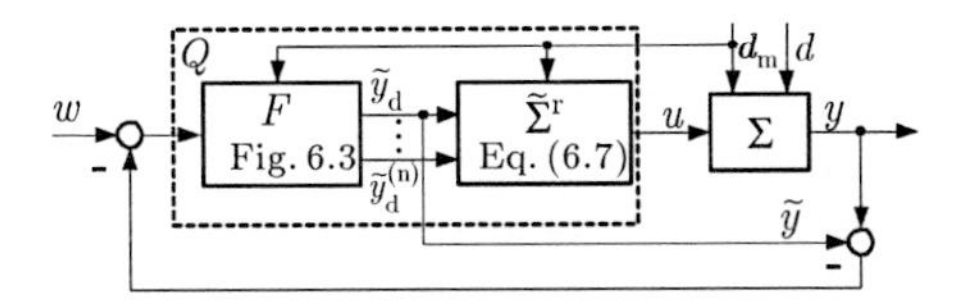

Fig. 6.4: Closed-loop control structure of the flatness-based IMC boost pressure controller. The calculations of the boundaries in Eq. (6.10) are not shown in this figure.

The complete flatness-based IMC controller can thus be written by (cf. Theorem 4.1)

$$Q: \quad \frac{1}{\lambda^2}\ddot{\tilde{y}}_\mathrm{d} + \frac{2}{\lambda}\dot{\tilde{y}}_\mathrm{d} + \tilde{y}_\mathrm{d} = \tilde{w} \quad \text{with } \ddot{\tilde{y}}_\mathrm{min} \leq \ddot{\tilde{y}}_\mathrm{d} \leq \ddot{\tilde{y}}_\mathrm{max} \text{ from Eq. (6.10)} \tag{6.11a}$$

$$u = \frac{\ddot{\tilde{y}}_\mathrm{d} - \varphi_3(\tilde{y}_\mathrm{d}, \dot{\tilde{y}}_\mathrm{d}, \boldsymbol{d}_\mathrm{m})}{\varphi_5(\tilde{y}_\mathrm{d}, \dot{\tilde{y}}_\mathrm{d}, \boldsymbol{d}_\mathrm{m})} \tag{6.11b}$$

where the differential equation (6.11a) presents the dynamics of the IMC filter F with the conditions to respect input constraints. The initial conditions of the filter are obtained from Eqns. (6.2) and (6.3) as

$$\begin{aligned} \tilde{y}_\mathrm{d}(0) &= x_1(0) \\ \dot{\tilde{y}}_\mathrm{d}(0) &= \frac{\varphi_4(x_2(0))}{k_7}\varphi_2(x_1(0), x_2(0)) - \frac{k_9}{k_7}k_6(\boldsymbol{d}_\mathrm{m}(0))x_2(0). \end{aligned} \tag{6.11c}$$

Equation (6.11b) is the right inverse $\tilde{\Sigma}^\mathrm{r}$.

According to the IMC properties, zero steady-state offset is to be expected. The closed-loop algorithm has no iterations and can be implemented online. Furthermore, the input constraints Eq. (6.9) will never be violated. This holds for arbitrary modelling errors and for arbitrary measured and unmeasured disturbances.

6.4 Test Bed Results

The IMC controller was tested on an engine test bed. The results portray the capabilities of the control concept. The parameter λ of the transfer function (6.8) of F is the degree-of-freedom of this IMC. It can be selected to be a compromise between performance and robustness of the closed-loop behaviour. In this case, the calibration of λ was performed manually

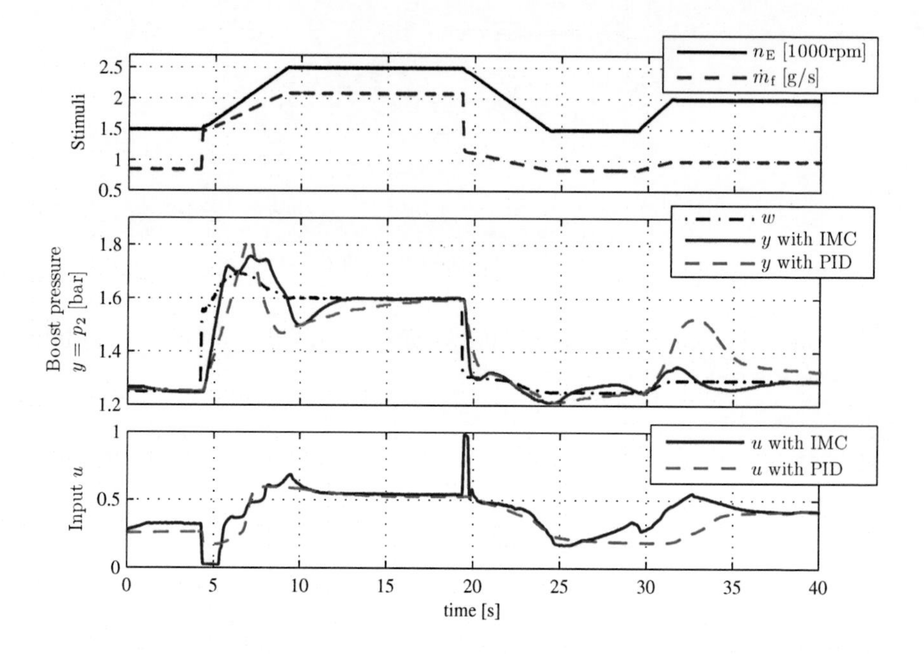

Fig. 6.5: Closed-loop performance on an engine test bed.

at the test bed which allowed to consider measurement noise and the performance criterions on the final plant. Most model parameters were identified off-line using manufacturer data and additional measurements. Only the characteristics of the vacuum-controlled diaphragm box, which drives the VNT position, was adjusted at the test bed. Calibration of the IMC was completed at the test bed within minutes. The input constraints are $0 \leq u(t) \leq 1$.

Figure 6.5 (showing time $0s \leq t \leq 40$s), Fig. 6.6 (showing time $200s \leq t \leq 290$s), and Fig. 6.7 (showing time $290s \leq t \leq 350$) portray the closed-loop tracking performance of the boost pressure p_2 at the test bed after calibration. The time intervals were chosen since they show the most significant differences between the gain-scheduled PID and the IMC controller. At all other times, the two controllers performed nearly identically.
The reference value w is generated according to the engine manufacturer specifications in dependence on the current engine speed n_E and fuel mass flow $\dot{m}_f$, which are portrayed in the first subplot of each figure. The second subplot shows a comparison of the output $y = p_2$ between the IMC

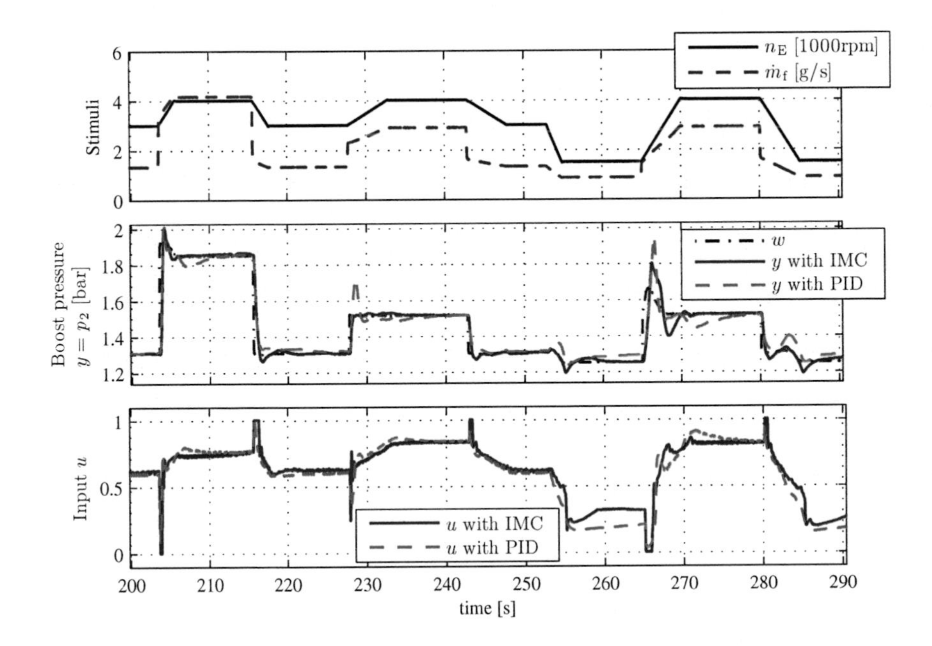

Fig. 6.6: Closed-loop performance on an engine test bed.

controller and the series production controller of that engine, labelled as PID. The series production controller has the structure shown in Fig. 1.3 and is a gain-scheduled PID with final calibration. Subplot three shows the comparison of the input signals of the IMC and the gain-scheduled PID controller.

One finds that the IMC controller is superior to the series production gain-scheduled PID controller considering its tracking behaviour. At times $t = 33$s, $t = 227$s, $t = 267$s and $t = 297$s the gain-scheduled PID controller shows more overshoot than the IMC controller. Less overshoot can also be observed, for example, at times $t = 265$s and $t = 328$s. The $\approx 40\%$ overshoot of the gain-scheduled PID controller at time $t = 33$s is not due to a change in the reference signal w but rather due to a change in engine speed n_E. One concludes that IMC compensates measured disturbances better (to which the engine speed n_E belongs) than the PID-type controller, which relies on the scheduling algorithm to deal with such influences.

In conclusion, the IMC tracks the reference signal better in many situ-

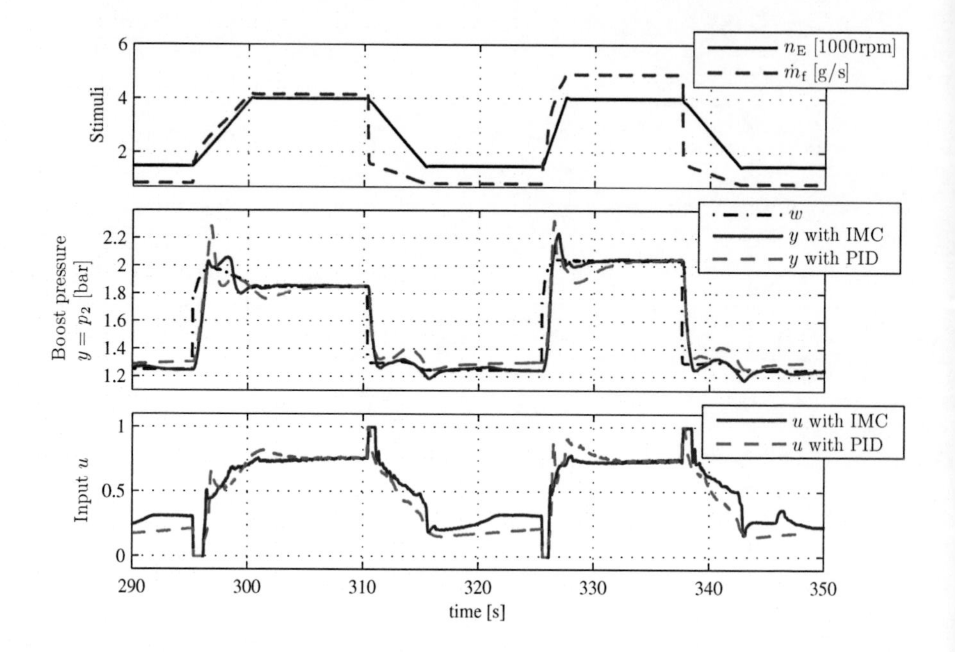

Fig. 6.7: Closed-loop performance on an engine test bed.

ations. As expected, the IMC controller achieves zero steady-state offset, despite significant modelling errors and disturbances, including unmeasured sensor noise. Thus, the IMC controller surpasses the series production controller in terms of performance with comparable high frequency input gain. The implementation effort of this nonlinear IMC controller is to be considered small. The controller itself has only two integrators (the IMC filter) to be implemented and the equations require only few and easily computed operations.

6.5 Summary

This chapter solves the pressure control problem of a one-stage turbocharged diesel engine using a variable-nozzle turbine. Since the simplified model of the air-system is flat, a flatness-based IMC is developed. The IMC filter was chosen to respect input constraints and the resulting controller was tested on a real engine at a test bed. The flatness-based IMC compares favourably to the current series production gain-scheduled PID controller in terms of performance and calibration effort. From an economical point of view, a nonlinear IMC is to be preferred over the typical PID controller design used in the automotive industry since it can be calibrated significantly faster, saving personnel cost. Moreover, an IMC uses less memory, since only the value of the physical parameters need to be stored (about thirty parameters) instead of about a dozen look-up tables with hundreds of values each. The computational effort is to be considered to be about in the same order of magnitude as the production PID.

7. CONTROL OF A TWO-STAGE TURBOCHARGED DIESEL ENGINE

This chapter develops an IMC pressure controller of a two-stage turbocharged air-system. Figure 7.1 shows the two-stage turbocharged air-system. The controller achieves tracking control of pressures in three pipes, namely the boost pressure, the exhaust back pressure and the inner turbines pressure, and is able to drive the plant into any physically possible state.

This control solution employs the basic IMC design procedure (cf. Section 3.5) where the model inverse is developed exploiting the input-affine plant model (cf. Section 4.2). The design is extended to handle input constraints (cf. Sections 5.4), to handle the singularity of the inverse of the plant model (cf. Section 5.2) and to incorporate measured disturbances (cf. Section 5.4).

This chapter presents the main practical contribution of this thesis. It is the first published control solution which is able use the full capabilities of this plant.

7.1 Function of a Two-Stage Turbocharged Diesel Engine

For the next generation of turbocharged diesel engines, current research is going towards the use of various combinations of turbines and compressors. One of the most promising combinations [3, 79, 90] is two-stage turbocharging. The motivation for this system is its higher possible boost pressure, reduced turbo lag[1], and lower boost threshold[2], in comparison to one-stage turbocharged engines.

---

[1] Turbo lag denotes the time the air-system needs to build the desired boost pressure.

[2] Boost threshold denotes the lowest engine speed n_{E} at which the air-system can build significant boost pressure.

The idea of charging an engine in various stages has existed since the early 20th century [105]. However, it has recently produced interest amongst car manufacturers for several reasons: The higher achievable boost pressures allow significant downsizing[3], have potential for significantly lower emissions through higher pressure ratios, and result in more "fun to drive".

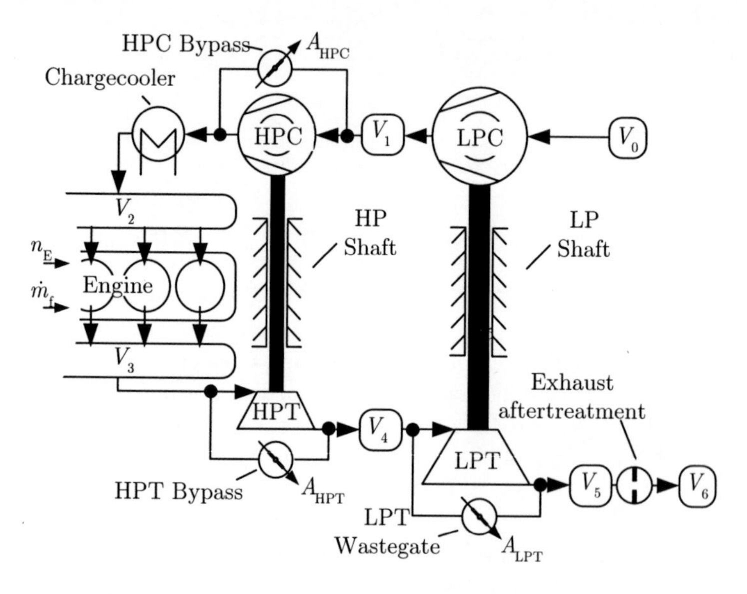

Fig. 7.1: Two-stage turbocharged engine.

In the concept of two-stage turbocharging (see Fig. 7.1), two turbochargers are placed in sequence. Each turbocharger is called a "stage". The turbocharger closest to the engine is the high-pressure (HP) stage and the low-pressure (LP) stage is positioned closer to the environment.

In the following, the air path through the system is described, starting with air from the environment V_0. The first compressor is the low-pressure compressor (LPC). The air is compressed for a second time in the high-pressure compressor (HPC). The HPC can be bypassed through an orifice with variable cross section area A_{HPC}. After combustion, the exhaust is first led over the high-pressure turbine (HPT), through the pipe V_4, and then over the low-pressure turbine (LPT). Both turbines can be bypassed

[3] The term "downsizing" refers to building engines with smaller displacement that still generate a high power output through forced induction.

by setting the cross section areas A_{HPT} and A_{LPT} accordingly. The cross section areas influence the amount of exhaust gas flowing through the turbines, which ultimately drive the compressors via the shafts.

The high pressure turbocharger is smaller and lighter and designed to operate at its optimum when the engine is at low speeds (i. e., under 2500rpm) enabling it to quickly deliver significant boost pressure from about 1300rpm. At higher engine speeds, the high pressure turbocharger reaches its saturation and is bypassed. In the range from about 2500rpm up to about 6000rpm, the low-pressure turbocharger is designed to deliver the necessary boost pressure.

Currently, only one manufacturer, namely BMW[4], offers a two-stage turbocharged engine. It is a three litres inline six cylider engine [18, 90]. However, all major car manufacturers are currently developing two-stage turbocharged engines, e. g., [38]. One of the major obstacles in introducing this new technology is its difficult control problem [3].

7.2 Control Problem

A two-stage turbocharged air-system (cf. Fig. 7.1) is structurally different from a one-stage air-system (cf. Fig. 6.1) in such that it has three inputs instead of one. Those are the bypass cross sections $A_{\mathrm{HPC}}, A_{\mathrm{HPT}}$ and A_{LPT}. The three inputs allow to pose requirements in addition to controlling the boost pressure p_2. In this work, the freedom given by the inputs is used to track reference values for the exhaust back pressure p_3 and the pressure between the turbines p_4. Figure 7.2 shows the input and output signals

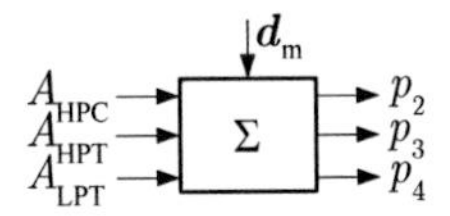

Fig. 7.2: Inputs and output signals of a two-stage turbocharged engine.

of a two-stage turbocharged engine.

Control Problem (Two-stage). The measured variable $\boldsymbol{y}$ is

$$\boldsymbol{y} = [p_2, p_3, p_4]^T. \tag{7.1}$$

[4] BMW named its two-stage turbocharged engines "twin charged". Colloquially, two-stage turbocharging is sometimes called "twin turbocharging". However, the expression "twin turbocharging" is not properly defined. It is used inconsistently for various (other) charging solutions and it is not used in scientific literature.

The available input consists of the cross section areas of the bypasses

$$\boldsymbol{u} = [A_{\text{HPC}}, A_{\text{HPT}}, A_{\text{LPT}}]^T. \tag{7.2}$$

Input constraints are as follows.

$$\boldsymbol{0} = \boldsymbol{u}_{\min} \leq \boldsymbol{u} \leq \boldsymbol{u}_{\max} \leq 12.5 \tag{7.3}$$

The requirements are summarised below.

- The output $\boldsymbol{y}$ should track the reference value $\boldsymbol{w}$ without steady-state offset.
- The effect of the measured disturbances $\boldsymbol{d}_{\text{m}}$ should be compensated.
- The closed-loop should be robustly stable.

The reason for the choice of the controlled variables is given in the following: The exhaust back pressure p_3 determines the force against which the exhaust gas needs to be pushed out. A low value results in better engine efficiency and a high value increases the engine braking power. Depending on the current driving condition, one is to be favoured over the other and, thus, this choice offers a straightforward influence on an important property of engine operation. Note that in the case of the one-stage turbocharged engine, the value of the exhaust back pressure p_3 was fully determined by the choice of the boost pressure p_2 (which is the controlled variable). The choice of the inner turbines pressure p_4 comes from the need of a third controlled variable for the control problem to be determined and since it yields a relative degree to the input u_3 of one. Moreover, this value offers the engineer a direct influence over which stage should provide how much thermal energy.

The raise in difficulty of controlling a two-stage compared to a one-stage turbocharged air-system is a structural change from a SISO to a MIMO control problem. It will also be shown below, that the two-stage turbocharged air-system becomes singular at some operating conditions which also presents an additional control challenge.

Model. The derivation of the model equations is given in Appendix A. For the sake of better readability, the equations of the resulting control-oriented model are repeated here. The six state variables of the reduced model of a two-stage turbocharged diesel engine are explained in the Appendix in Tab. A.1 and Tab. A.2 introduces the parameters.

The reduced-order, input-affine model $\widetilde{\Sigma}$, with which the IMC controller is developed, is given by

$$\widetilde{\Sigma}: \quad \dot{\boldsymbol{x}} = \boldsymbol{f}(\boldsymbol{x}, \boldsymbol{d}_{\mathrm{m}}) + \boldsymbol{G}(\boldsymbol{x}, \boldsymbol{d}_{\mathrm{m}})\boldsymbol{u}, \quad \boldsymbol{x}(0) = \boldsymbol{x}_0, \quad \boldsymbol{x} \in \mathbb{R}^6 \ \boldsymbol{u} \in \mathbb{R}^3 \tag{7.4a}$$

$$\tilde{\boldsymbol{y}} = \boldsymbol{h}(\boldsymbol{x}) = \begin{bmatrix} x_1, & x_2, & x_3 \end{bmatrix}^T, \quad \boldsymbol{y} \in \mathbb{R}^3, \tag{7.4b}$$

with model output $\tilde{\boldsymbol{y}}$. The vector field $\boldsymbol{f}$ is defined as

$$\boldsymbol{f}(\boldsymbol{x}, \boldsymbol{d}_{\mathrm{m}}) = \begin{bmatrix} \frac{\kappa_2 - 1}{V_2}\left(\dot{H}_{\mathrm{HPC,out}} + \dot{H}_{\mathrm{E,in}}\right) \\ \frac{\kappa_3 - 1}{V_3}\left(\dot{H}_{\mathrm{E,out}} + \dot{H}_{\mathrm{HPT,in}}\right) \\ \frac{\kappa_4 - 1}{V_4}\left(\dot{H}_{\mathrm{HPT,out}} + \dot{H}_{\mathrm{LPT,in}}\right) \\ \dot{m}_{\mathrm{HPT,out}} + \dot{m}_{\mathrm{LPT,in}} \\ \frac{1}{J_{\mathrm{LP}}\omega_{\mathrm{LP}}}\left(P_{\mathrm{LPT}} - P_{\mathrm{LPC}} - d_{\mathrm{LP}}\omega_{\mathrm{LP}}^2\right) \\ \frac{1}{J_{\mathrm{HP}}\omega_{\mathrm{HP}}}\left(P_{\mathrm{HPT}} - P_{\mathrm{HPC}} - d_{\mathrm{HP}}\omega_{\mathrm{HP}}^2\right) \end{bmatrix} \tag{7.4c}$$

and the matrix $\boldsymbol{G}$ has the following structure:

$$\boldsymbol{G}(\boldsymbol{x}, \boldsymbol{d}_{\mathrm{m}}) = \begin{bmatrix} g_{11}(\boldsymbol{x}, \boldsymbol{d}_{\mathrm{m}}) & 0 & 0 \\ 0 & g_{22}(\boldsymbol{x}, \boldsymbol{d}_{\mathrm{m}}) & 0 \\ 0 & g_{32}(\boldsymbol{x}, \boldsymbol{d}_{\mathrm{m}}) & g_{33}(\boldsymbol{x}, \boldsymbol{d}_{\mathrm{m}}) \\ 0 & g_{42}(\boldsymbol{x}, \boldsymbol{d}_{\mathrm{m}}) & g_{43}(\boldsymbol{x}, \boldsymbol{d}_{\mathrm{m}}) \\ 0 & 0 & 0 \\ 0 & 0 & 0 \end{bmatrix} \cdot 1/\mathrm{m}^2. \tag{7.4d}$$

The description of the components of G is given in the Appendix in Eq. (A.45e).

7.3 Literature Review

The choice of controlled variables given above, does not reflect any current demand from a car manufacturer. Since this charging solution is a dramatic structural change (from one to three controlled variables), car manufacturers and suppliers find it difficult to fully exploit its possibilities. Rather, it is desired to maintain the control problem of the one-stage turbocharged engine (control the boost pressure with one input variable) for which a solution has grown over the last two decades.

There are few articles concerning the control of two-stage turbocharged engines, namely [96][5] and [90][6]. The control problem solved by existing

[5] Developed by the supplier ERFI

[6] Developed by the car manufacturer BMW

solutions consists of only one controlled variable, namely the boost pressure p_2. This, however, yields an underdetermined control problem, since a specific boost pressure p_2 can be obtained by infinitely many choices of the three control inputs. This dilemma is evaded by *switching between three SISO controllers* which all control the boost pressure p_2 but with different input variables. Thus, in dependence upon the current operating point, an overlying logic sets two of the three inputs at some constant value (either fully open or fully closed) and uses the remaining input to control the boost pressure p_2. Which inputs are kept constant and which is used to control the boost pressure p_2 depends on thresholds of either the current values of engine speed n_{E} and fuel mass flow $\dot{m}_{\mathrm{f}}$ [90] or the value of the estimated turbocharger speeds $n_{\mathrm{LP}}, n_{\mathrm{HP}}$ and the boost pressure p_2 [96].

The cited solutions lack the flexibility of a MIMO controller as it is proposed in this chapter, since it has the potential to drive the plant through all its physically possible behaviour simply by changing the reference values. For example, some exhaust gas aftertreatment methods require a boost pressure below ambient pressure. A two-stage turbocharged engine has the potential to do this. However, the current solutions are not capable of driving the boost pressure to such a low value since the overlying switching logic opens and closes two of the three bypasses for normal driving, prohibiting to reach the necessary low boost pressure p_2 with only the third input. Thus, the existing scheduling algorithms would need to be enhanced by additional operation modes (cf. Section 1.3.3), which would lead to a more complex switching logic. In the case of a MIMO controller, the necessary pressures can be reached simply by changing the reference values.

In the case of [90], the solution was first commercially available in the 2006 BMW 535d. The control law is essentially a PID controller of the same structure as shown in Chapter 1.3 with the extension that it can change the choice of input variable. It works well with a large enough adaptation algorithm and is not based on a methodological design.

The solution of [96] includes an algebraic feedforward control based on physical laws and includes a PID feedback controller. Nominal stability cannot be guaranteed. This approach can also be put into the typical structure of automotive controller shown in Chapter 1.3 with the only difference of switching the choice of input and computing the *static* feedforward control input via physical laws (instead of measuring it at the test bed).

In conclusion, the control problem is difficult and unsolved in the sense that no control algorithm exists which is able to drive the system into all physically possible states. The existing solutions are still based on an ad hoc approach, are not able to take advantage of the full potentials of the plant, offer no guarantees concerning nominal stability, and require a high calibration effort.

In the following, a MIMO IMC controller is developed which is able to drive the plant into any physically possible state, guarantees nominal and robust stability and respects the input constraints.

7.4 Nonlinear IMC of the Air-System

7.4.1 Model Inverse

The model (7.4) is input-affine and allows to create an inverse using the I/O normal form as described in Section 4.2. Here, the MIMO case needs to be considered as discussed in Section 5.5.

As in the SISO case, the transformation of the system equations into the I/O normal form begins with repetitive differentiations of the individual outputs $\tilde{y}_i$ in Eq. (7.4b) with respect to time and the substitution of the elements by the state equations Eq. (7.4a) until an input u_j appears explicitly. One finds

$$\begin{aligned}\begin{bmatrix}\dot{\tilde{y}}_1\\ \dot{\tilde{y}}_2\\ \dot{\tilde{y}}_3\end{bmatrix} &= \overbrace{\begin{bmatrix}f_1(\boldsymbol{x},\boldsymbol{d}_\mathrm{m})\\ f_2(\boldsymbol{x},\boldsymbol{d}_\mathrm{m})\\ f_3(\boldsymbol{x},\boldsymbol{d}_\mathrm{m})\end{bmatrix}}^{\boldsymbol{a}(\boldsymbol{x},\boldsymbol{d}_\mathrm{m})} \\ &\quad + \underbrace{\begin{bmatrix}g_{11}(\boldsymbol{x},\boldsymbol{d}_\mathrm{m}) & 0 & 0\\ 0 & g_{22}(\boldsymbol{x},\boldsymbol{d}_\mathrm{m}) & 0\\ 0 & g_{32}(\boldsymbol{x},\boldsymbol{d}_\mathrm{m}) & g_{33}(\boldsymbol{x},\boldsymbol{d}_\mathrm{m})\end{bmatrix}}_{\boldsymbol{B}(\boldsymbol{x},\boldsymbol{d}_\mathrm{m})}\begin{bmatrix}u_1\\ u_2\\ u_3\end{bmatrix},\end{aligned} \tag{7.5}$$

where the first derivative of each output component $\tilde{y}_i$ already contains an input component u_j. Hence, the relative degree with respect to each input is one. Each input affects a different output with this relative degree[7]. Since only one derivative is necessary, no differentiations of the measured disturbance $\boldsymbol{d}_\mathrm{m}$ occur (cf. Section 5.4). The matrix $\boldsymbol{B}$ is called the decoupling matrix.

[7] Note that y_3 was chosen such that each input has a relative degree of one.

With Eq. (7.5), a transformation $\boldsymbol{\Phi}(\boldsymbol{x})$ of the system $\widetilde{\Sigma}$ into I/O normal form (see e. g., [51]) is obtained by

$$\begin{bmatrix} \tilde{\boldsymbol{y}} \\ \boldsymbol{\eta} \end{bmatrix} = \boldsymbol{\Phi}(\boldsymbol{x}) = \begin{bmatrix} x_1 & x_2 & x_3 & x_4 & x_5 & x_6 \end{bmatrix}^T . \tag{7.6}$$

Thus, the system $\widetilde{\Sigma}$ is *already* in I/O normal form. The transformed states are $\tilde{\boldsymbol{y}} = \begin{bmatrix} x_1 & x_2 & x_3 \end{bmatrix}^T$ and $\boldsymbol{\eta} = \begin{bmatrix} x_4 & x_5 & x_6 \end{bmatrix}^T$. The states $\tilde{\boldsymbol{y}}$ will be controlled directly (since they are also outputs) whereas the states $\boldsymbol{\eta}$ are the states of the internal dynamics. With the demand

$$\tilde{\boldsymbol{y}} \overset{!}{=} \boldsymbol{y}_{\mathrm{d}} \tag{7.7}$$

the right inverse $\widetilde{\Sigma}^{\mathrm{r}}$ is obtained by solving Eq. (7.5) for the input $\boldsymbol{u}$ with (7.6). One gets

$$\boldsymbol{u} = \boldsymbol{B}^{-1}\left(\boldsymbol{y}_{\mathrm{d}}, \boldsymbol{\eta}, \boldsymbol{d}_{\mathrm{m}}\right) \cdot \left(\dot{\boldsymbol{y}}_{\mathrm{d}} - \boldsymbol{a}\left(\boldsymbol{y}_{\mathrm{d}}, \boldsymbol{\eta}, \boldsymbol{d}_{\mathrm{m}}\right)\right) . \tag{7.8}$$

In Section 4.2 it is proposed to obtain the states $\boldsymbol{\eta}$ of the internal dynamics by integrating their differential equation

$$\begin{aligned} \dot{\boldsymbol{\eta}} &= \begin{bmatrix} f_4(\boldsymbol{y}_{\mathrm{d}}, \boldsymbol{\eta}, \boldsymbol{d}_{\mathrm{m}}) \\ f_5(\boldsymbol{y}_{\mathrm{d}}, \boldsymbol{\eta}, \boldsymbol{d}_{\mathrm{m}}) \\ f_6(\boldsymbol{y}_{\mathrm{d}}, \boldsymbol{\eta}, \boldsymbol{d}_{\mathrm{m}}) \end{bmatrix} + \begin{bmatrix} 0 & g_{42}\left(\boldsymbol{y}_{\mathrm{d}}, \boldsymbol{\eta}, \boldsymbol{d}_{\mathrm{m}}\right) & g_{43}\left(\boldsymbol{y}_{\mathrm{d}}, \boldsymbol{\eta}, \boldsymbol{d}_{\mathrm{m}}\right) \\ 0 & 0 & 0 \\ 0 & 0 & 0 \end{bmatrix} \boldsymbol{u} \\ \boldsymbol{\eta}(0) &= \begin{bmatrix} x_4(0) & x_5(0) & x_6(0) \end{bmatrix}^T \end{aligned} \tag{7.9}$$

numerically. Simulations show that the solution of the internal dynamics is stable.

According to Theorem 4.2, the input $\boldsymbol{u}$ can be calculated by the inverse $\widetilde{\Sigma}^{\mathrm{r}}$ shown in Fig. 7.3. Note that the measured disturbance $\boldsymbol{d}_{\mathrm{m}}$ also enters Eqns. (7.9) and (7.8) but is omitted in Fig. 7.3 for clarity.

In short, Eq. (7.8) together with the solution of Eq. (7.9) represent the model inverse $\widetilde{\Sigma}^{\mathrm{r}}$. However, the decoupling matrix $\boldsymbol{B}$ in Eq. (7.5) must be regular, otherwise the inverse would not be defined as the relative degree might be lost.

7.4.2 Singularity of the Model Inverse of a Two-Stage Turbocharged Engine

The inputs $\boldsymbol{u}$ are the cross-sections of the bypasses (cf. Fig. 7.1) $\boldsymbol{u} = [A_{\mathrm{HPC}}, A_{\mathrm{HPT}}, A_{\mathrm{LPT}}]^T$. If the pressures before and after a bypass are

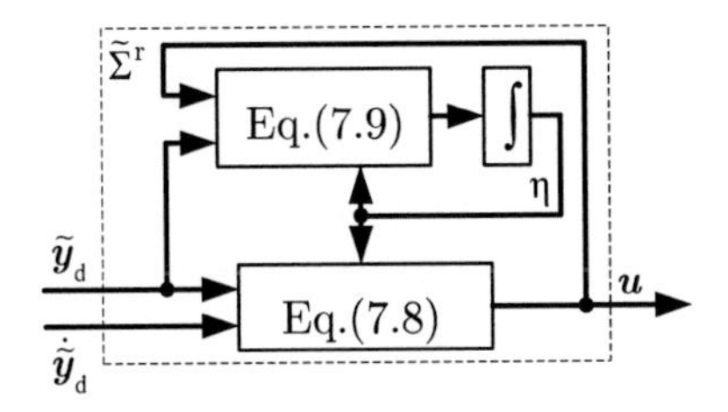

Fig. 7.3: Structure of the right inverse $\widetilde{\Sigma}^{\mathrm{r}}$ of the model (7.4) of the two-stage turbocharged air-system.

equal then the mass flow through the bypass vanishes independently of its cross-section. Then, the input $\boldsymbol{u}$ cannot influence the system which results in a singularity (cf. Section 5.2).

The mathematical equivalence of this problem is that the decoupling matrix $\boldsymbol{B}$ loses its regularity. The inverse of the matrix $\boldsymbol{B}$ in Eq. (7.8)

$$\boldsymbol{B}^{-1} = \begin{bmatrix} 1/g_{11} & 0 & 0 \\ 0 & 1/g_{22} & 0 \\ 0 & -g_{32}/(g_{22}g_{33}) & 1/g_{33} \end{bmatrix} \tag{7.10}$$

is only defined if all denominators in Eq. (7.10) are non-zero:

$$g_{11}, g_{22}, g_{33} \neq 0. \tag{7.11}$$

For the elements g_{22} and g_{33} (cf. Eq. (A.45e)) this is guaranteed by the physical mode of operation: All pressures from the engine into the environment V_6 decrease and, therefore, there would always be positive fluid flow and therewith positive enthalpy flow through the turbine bypasses if they had unitary cross section areas.

However, the term g_{11} can become zero during normal driving conditions. It can be interpreted as a the part of the enthalpy flow that is responsible for building the boost pressure (cf. Eqns. (A.3) and (A.45e)). This happens if the pressure p_1 between the compressors is equal to the boost pressure p_2.

$$p_1 = p_2 \Leftrightarrow g_{11} = 0. \tag{7.12}$$

This is the case, for example, when the high-pressure turbocharger is bypassed and the required boost pressure is delivered by the low-pressure compressor. Then, a pressure equalisation between the pipes V_1 and V_2 (cf. Fig. 7.1) can take place and no fluid flows through the HPC bypass, independently of its cross section area. Thus, at $p_1 = p_2$ the feedforward controller is not defined since the plant becomes singular.

As discussed in Section 5.2, a loss in relative degree or invertibility can be accommodated by an IMC filter which also respects input constraints. Such a filter is introduced in the following section.

7.4.3 IMC Filter

In Section 5.5 it was established that the IMC filter for MIMO systems is a diagonal operator. Here, it is proposed to design each diagonal element F_i (with $i = 1, 2, 3$) as

$$\frac{\tilde{y}_{\mathrm{d}i}(s)}{\tilde{w}_i(s)} = \frac{1}{s/\lambda_i + 1}. \tag{7.13}$$

Each has a pole at $-\lambda_i$. The series of the SVFs with the right inverse shown in Fig. 7.3 yields a realisable MIMO feedforward controller.

In Section 5.2 it was established that the introduction of input constraints also handles model singularities. Here, it will be shown, using plant-specific explanations, that the singularity at $p_1 = p_2$ is handled by the input constraints. Equation (7.5) presents an algebraic relationship between the inputs $\boldsymbol{u}$ and the highest derivative $\dot{\tilde{\boldsymbol{y}}}$. This relationship also holds true, if some (or all) of the inputs saturate and yields the respective maximal and minimal speeds $\dot{\tilde{y}}_{i\max}, \dot{\tilde{y}}_{i\min}$.

$$\begin{aligned}
\dot{\tilde{y}}_{1\max} &= f_1 + \max\left(g_{11}u_{1,\max}, g_{11}u_{1,\min}\right) \\
\dot{\tilde{y}}_{1\min} &= f_1 + \min\left(g_{11}u_{1,\max}, g_{11}u_{1,\min}\right) \qquad (7.14a) \\
\dot{\tilde{y}}_{2\max} &= f_2 + \max\left(g_{22}u_{2,\max}, g_{22}u_{2,\min}\right) \\
\dot{\tilde{y}}_{2\min} &= f_2 + \min\left(g_{22}u_{2,\max}, g_{22}u_{2,\min}\right) \qquad (7.14b) \\
\dot{\tilde{y}}_{3\max} &= f_3 + g_{32}u_2 + \max(g_{33}u_{3,\max}, g_{33}u_{3,\min}) \\
\dot{\tilde{y}}_{3\min} &= f_3 + g_{32}u_2 + \min(g_{33}u_{3,\max}, g_{33}u_{3,\min}). \qquad (7.14c)
\end{aligned}$$

If the IMC filter ensures that the speeds $\dot{\tilde{y}}_{\mathrm{d}i}$ are restricted to

$$\dot{\tilde{y}}_{i\min} \leq \dot{\tilde{y}}_{\mathrm{d}i} \leq \dot{\tilde{y}}_{i\max} \tag{7.15}$$

then the input $\boldsymbol{u}$ obtained by Eq. (7.8) never violates its limitations (7.3). The limited IMC filter is shown in Fig. 7.4.

Finally, the following corollary says that the right inverse is defined despite moving through the singularity if the above mentioned filter is used:

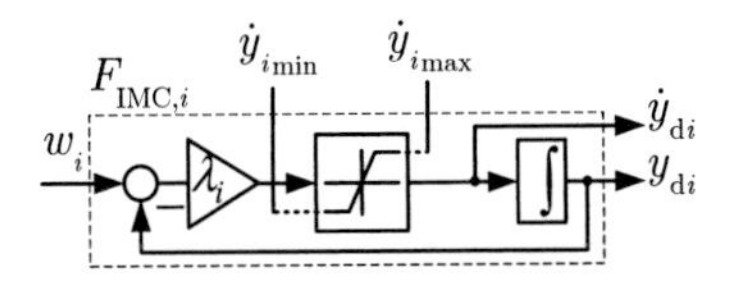

Fig. 7.4: Nonlinear IMC filter $F_{\text{IMC},i}$ with saturation.

Corollary 7.1. *If the IMC filter from Fig. 7.4 is used with the limits described by Eq.* (7.14) *then a feedforward controller using Eq.* (7.8) *is globally defined despite the singularity of the matrix* $\boldsymbol{B}^{-1}$ *from Eq.* (7.10).

Proof. As the system approaches the singular point $g_{11} \to 0$ the input u_1 will necessarily saturate at either boundary. Then, using the IMC filter from Fig. 7.4 with Eq. (7.14) one finds

$$\dot{y}_{\text{d}1} = f_1 + g_{11} u_{1\max} \quad \text{or} \quad \dot{y}_{\text{d}1} = f_1 + g_{11} u_{1\min} \tag{7.16}$$

and in general Eq. (7.8) yields for the first input u_1

$$u_1 = \frac{\dot{y}_{\text{d}1} - f_1}{g_{11}}. \tag{7.17}$$

Using Eq. (7.16) in Eq. (7.17) results in

$$u_1 = \frac{g_{11}}{g_{11}} u_{1\max} \quad \text{or} \quad u_1 = \frac{g_{11}}{g_{11}} u_{1\min}. \tag{7.18}$$

Taking the limit of Eq. (7.18) as $g_{11} \to 0$ (using L'Hospital's rule)

$$\lim_{g_{11}\to 0} u_1 = u_{1\max} \quad \text{or} \quad \lim_{g_{11}\to 0} u_1 = u_{1\min} \tag{7.19}$$

shows that the feedforward control from Eq. (7.8) yields a valid input which will be placed at either boundary . □

From a physical point of view, the explanation why the IMC filter from Fig. 7.4 results in a valid input despite the rank deficiency is the following: Since the HPC bypass has no influence at the singular point $p_1 = p_2$, the model can only travel through it in a certain fashion (namely with the exact rate $\dot{\tilde{y}}_1 = f_1(\boldsymbol{x}, \boldsymbol{d}_\text{m})$). The IMC filter ensures that the trajectory $\boldsymbol{y}_\text{d}$, which is generated from the reference signal $\boldsymbol{w}$, can be achieved by the system exactly (see Eq. (7.7)) *with* permissible inputs. Thereby, it automatically leads the system through the singularity with the prescribed rate.

7.4.4 Complete IMC Law

If the non-realisable (but perfect) feedforward controller $\widetilde{\Sigma}^{\mathrm{r}}$ from Fig. 7.3 is "padded" with an IMC filter $F_{\mathrm{IMC}i}$ from Fig. 7.4 for each reference signal w_i then a realisable feedforward controller Q results, which also respects the input constraints (7.3). The resulting structure of this feedforward controller is shown in Fig. 7.5. The IMC controller Q for a two-stage

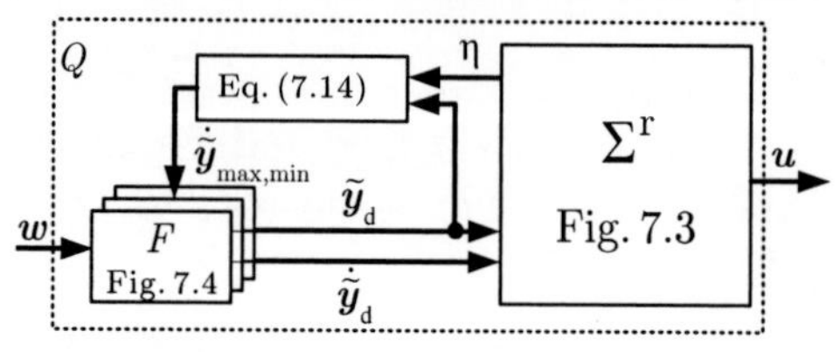

Fig. 7.5: IMC controller Q for a two-stage turbocharged air-system.

turbocharged air-system can thus be written as:

$$\frac{1}{\lambda_i}\dot{\tilde{y}}_{\mathrm{d}i} + \tilde{y}_{\mathrm{d}i} = \tilde{w}_i \quad \text{with } i = 1, 2, 3$$
$$\text{with } \tilde{y}_{i\min} \leq \dot{\tilde{y}}_{\mathrm{d}i} \leq \tilde{y}_{i\max} \text{ from Eq. (7.14)} \tag{7.20a}$$

$$\dot{\boldsymbol{\eta}} = \begin{bmatrix} f_4(\tilde{\boldsymbol{y}}_{\mathrm{d}}, \boldsymbol{\eta}, \boldsymbol{d}_{\mathrm{m}}) \\ f_5(\tilde{\boldsymbol{y}}_{\mathrm{d}}, \boldsymbol{\eta}, \boldsymbol{d}_{\mathrm{m}}) \\ f_6(\tilde{\boldsymbol{y}}_{\mathrm{d}}, \boldsymbol{\eta}, \boldsymbol{d}_{\mathrm{m}}) \end{bmatrix} + \begin{bmatrix} 0 & g_{42}(\tilde{\boldsymbol{y}}_{\mathrm{d}}, \boldsymbol{\eta}, \boldsymbol{d}_{\mathrm{m}}) & g_{43}(\tilde{\boldsymbol{y}}_{\mathrm{d}}, \boldsymbol{\eta}, \boldsymbol{d}_{\mathrm{m}}) \\ 0 & 0 & 0 \\ 0 & 0 & 0 \end{bmatrix} \boldsymbol{u} \tag{7.20b}$$
$$\boldsymbol{u} = \boldsymbol{B}^{-1}(\tilde{\boldsymbol{y}}_{\mathrm{d}}, \boldsymbol{\eta}, \boldsymbol{d}_{\mathrm{m}}) \cdot \left(\dot{\tilde{\boldsymbol{y}}}_{\mathrm{d}} - \boldsymbol{a}(\tilde{\boldsymbol{y}}_{\mathrm{d}}, \boldsymbol{\eta}, \boldsymbol{d}_{\mathrm{m}})\right)$$

where Eq. (7.20a) presents the dynamics of the MIMO filter F, that respects the input constraints. Equation (7.20b) presents the right inverse of the model with internal dynamics and control law. The initial condition for the filter dynamics and the internal dynamics are

$$\tilde{\boldsymbol{y}}_{\mathrm{d}}(0) = \left[x_1(0), \quad x_2(0), \quad x_3(0)\right]^T \tag{7.20c}$$
$$\boldsymbol{\eta}(0) = \left[x_4(0), \quad x_5(0), \quad x_6(0)\right]^T. \tag{7.20d}$$

This controller is containable in a standard OCU: It consists of six integrators (three integrators in the MIMO IMC filter and three integrators in the internal dynamics). The controller algorithm has to compute some powers and square roots on top of the usual additions and subtractions. It does not contain any computationally intense on-line operations, like matrix inversions.

The calibration of the controller concerns finding the model parameters and the three poles λ_i of the MIMO IMC filter F. The model parameters can be determined using the algorithm given in Section A.4 in about two hours. The filter poles can be calibrated within a matter of minutes.

In conclusion, this IMC is a feasible automotive controller, since it is implementable in a standard OCU and offers dedicated calibration parameters.

7.4.5 Robust Stability Analysis

This section is concerned with showing that the above designed nonlinear IMC controller for the two-stage air-system provides robust stability. A major issue in a stability analysis for industrial problems is that an upper boundary $\bar{\Delta}$ of unstructured uncertainties Δ is almost never given a priori. Thus, the first part of this analysis is focused on determining such an upper boundary $\bar{\Delta}$ for the two-stage turbocharged engine. In the second part, the actual analysis will take place.

Finding an upper boundary $\bar{\Delta}$

The procedure as performed in the following should be interpreted as a "hands-on" approach which can be performed in industry and should give the engineer a rough guess where the uncertainties lie. It should not be regarded as mathematically sufficient.

Main idea. The main idea is based on the observation presented in Section 3.5.3, namely, that the stability analysis of an IMC loop can be performed on the feedback loop in Fig. 3.10, where F is a linear system. In this analysis, a linear approximation of the uncertainty Δ is gained which is then used for a linear stability analysis.

The following deals with the sufficient conditions of stability given in Eqns. (3.40)- (3.43). Since the uncertainty Δ is a nonlinear system, its gain $g(\Delta)$ cannot be computed even if Δ was known. To this end, it is proposed to approximate the nonlinear uncertainty Δ by a linear (and Laplace transformed) approximation $\Delta(s)$. Using a linear approximation of the uncertainty Δ is reasonable since the plant model $\widetilde{\Sigma}$ itself should be chosen to contain the main nonlinearities of the plant.

In the following, a linear approximation (and Laplace transform) $\Delta(s)$ of the nonlinear uncertainty Δ is obtained. The closed-loop robustness analysis is then performed using the linear approximation $\Delta(s)$. For this procedure to be acceptable, the relationship

$$g(\Delta(s)) \geq g(\Delta) \tag{7.21}$$

must be assumed to hold. Hence, this section presents a method to obtain a linear approximation $\Delta(s)$ of the nonlinear model uncertainty Δ. The result of the following procedure is an upper boundary $\bar{\Delta}(\omega)$ which is then assumed to be *given* for the robustness analysis.

Procedure. First, a set of models $\mathcal{M}$ is generated from the nominal model $\widetilde{\Sigma}$ by varying a vector $\boldsymbol{p}$ of significant parameters of $\widetilde{\Sigma}$ within given boundaries:

$$\mathcal{M} = \left\{ \widetilde{\Sigma}_{\mathrm{p}} \mid \boldsymbol{p}_{\min} \leq \boldsymbol{p} \leq \boldsymbol{p}_{\max} \right\} \tag{7.22}$$

The expression $\widetilde{\Sigma}_{\mathrm{p}}$ signifies a model $\widetilde{\Sigma}$ with the parameterization $\boldsymbol{p}$. Thus, $\mathcal{M}$ represents a set of nonlinear models of two-stage turbocharged engines.

The open-loop behaviour Σ_0 of the generalised IMC structure in Fig. 3.4 is depicted in Fig. 7.6.

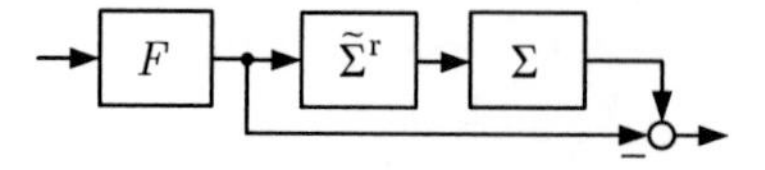

Fig. 7.6: Open-loop structure $\Sigma_0 = F \circ \left(\Sigma \widetilde{\Sigma}^{\mathrm{r}} - I \right)$ of an IMC controller $Q = \Sigma^{\mathrm{r}} F$.

It is proposed to linearise Σ_0 numerically around all operating points o while Σ is substituted by all models $\Sigma \rightarrow \widetilde{\Sigma}(\boldsymbol{p}) \in \mathcal{M}$. This is a required intermediate step since it incorporates the effect of the model inverse in obtaining the model uncertainty $\Delta(s)$. With Eq. (3.38) and

$$F(s) = \begin{bmatrix} \frac{1}{s/\lambda_1+1} & 0 & 0 \\ 0 & \frac{1}{s/\lambda_2+1} & 0 \\ 0 & 0 & \frac{1}{s/\lambda_3+1} \end{bmatrix} \tag{7.23}$$

each linearisation with model parameterization $\boldsymbol{p}$ and operating point o will give linear MIMO uncertainties $\Delta_{\mathrm{p},o}(s)$ by

$$\Delta_{\mathrm{p},o}(s) = G_{0_{\mathrm{p},o}}(s) F^{-1}(s). \tag{7.24}$$

Finally, an upper boundary $\bar{\Delta}(\omega)$ can be found by

$$\bar{\Delta}(\omega) = \max_{\omega}|\Delta_{\mathrm{p},o}(j\omega)|, \quad \forall\, \boldsymbol{p},\, o \tag{7.25}$$

where max $|\cdot|$ denotes an element-wise maximum of the magnitude of the transfer function. Thus, $\bar{\Delta}(\omega)$ is a matrix containing the function of the amplitude over frequency of each element. The elements at any frequency are equal to the largest amplitude of that element at that frequency of all obtained uncertainties $\Delta_{\mathrm{p},o}(s)$.

Figure 7.7 shows the result of the open-loop linearisations of the IMC controller for the two-stage turbocharged engine with modelling errors in the loop. The linearisations concern the feedforward control structure as shown in Fig. 7.6. Therewith, a linear upper boundary on the MIMO

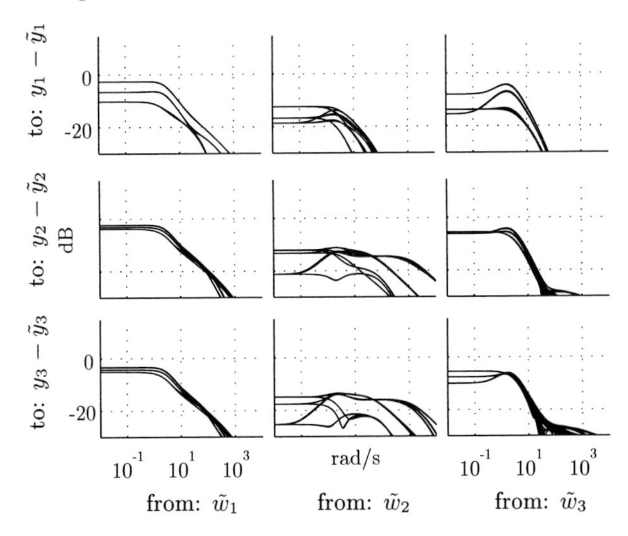

Fig. 7.7: Comparison of the amplitudes of the linearised models $|\Sigma_{0p,o}(j\omega)|$ of the open-loop structure (cf. Fig. 7.6).

output uncertainties $\bar{\Delta}(\omega)$ is determined which is shown in Fig. 7.8. An evaluation of the upper boundary $\bar{\Delta}(\omega)$ in Fig. 7.8 shows that some of its elements have a significant high frequency gain of about 20.

Stability analysis

It is assumed, that

$$g(\Delta) \leq g(\Delta_{p,o}(s)), \quad \forall p, o \tag{7.26}$$

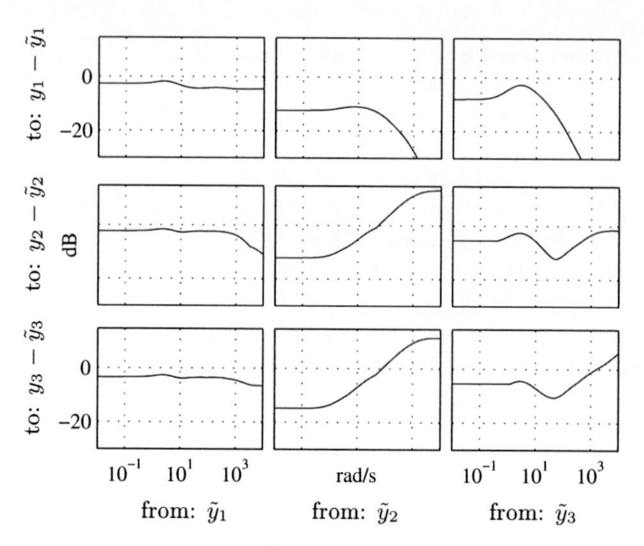

Fig. 7.8: Upper boundary $\bar{\Delta}(\omega)$ of the linearised model uncertainties derived from parameter variations.

holds; that is: The gain of the actual nonlinear uncertainties Δ is smaller or equal to the gain of the linearisation obtained with the procedure above. Additionally, the determination of the upper boundary $\bar{\Delta}(\omega)$ can in reality only be performed for a finite number of parameter variations and finite number of operating points. Nevertheless, it is assumed that such an approximation will give a representative upper boundary $\bar{\Delta}(\omega)$.

Finally, the maximum singular value $\sigma_{\max}$ can be computed for $\bar{\Delta}(j\omega)|F(j\omega)|$ for all frequencies ω. The result is displayed in Fig. 7.9. A visual inspection of $\sigma_{\max}$ in Fig. 7.9 shows that it is smaller than one for all frequencies ω. According to [64, 87], this implies robust stability of the closed-loop. The high frequency gain of the uncertainties is compensated by the filter F which can now be interpreted as ensuring robust stability at those frequencies.

Since model parameters were varied within manufacturing tolerances and the model represents the plant well, this presented approach yields sufficient confidence for the control engineer that the nonlinear IMC will successfully control the two-stage turbocharged diesel engine, despite manufacturing tolerances and ageing.

However, it is important to note that this approach is conservative. Simulations have shown stability for the closed-loop even for norms well

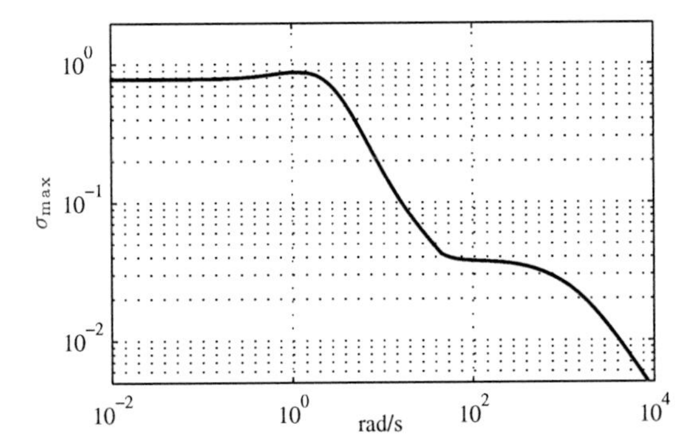

Fig. 7.9: Maximum singular value σ_{max} of $\bar{\Delta}(\omega)|F(j\omega)|$ over all frequencies ω.

beyond one. In simulations, there was no parameter variation which led to instability.

7.5 Simulation Results

Simulation results are presented in this section. It can be assumed that the simulation results are representative of a test bed experiment, since

- the IMC of the one-stage turbocharged air-system showed that simulation results are a close match to what is observed at the test bed and
- the model quality of the two-stage turbocharged air-system is excellent in both transient and steady-state behaviour.

The controlled plant is the original model with twelve states. Thus, modelling errors are present. The results are shown in Fig. 7.10. Engine speed n_{E} and fuel mass flow $\dot{m}_{\mathrm{f}}$ change during the simulation run. The engine speed n_{E} ramps from 3000rpm to 5500rpm in five seconds (from time 1s to 6s) under full throttle. At time 6s, acceleration ends and the engine proceeds in constant highway driving. Thus, until time 11s, a medium fuel flow and medium boost pressure is required. During this time, the boost pressure is delivered mainly by the LPC. From time 11s until 16s, the capabilities of this controller are displayed by demanding a boost pressure below ambient pressure. The controller achieves this by closing the HPC bypass and opening both turbine bypasses. This behaviour cannot

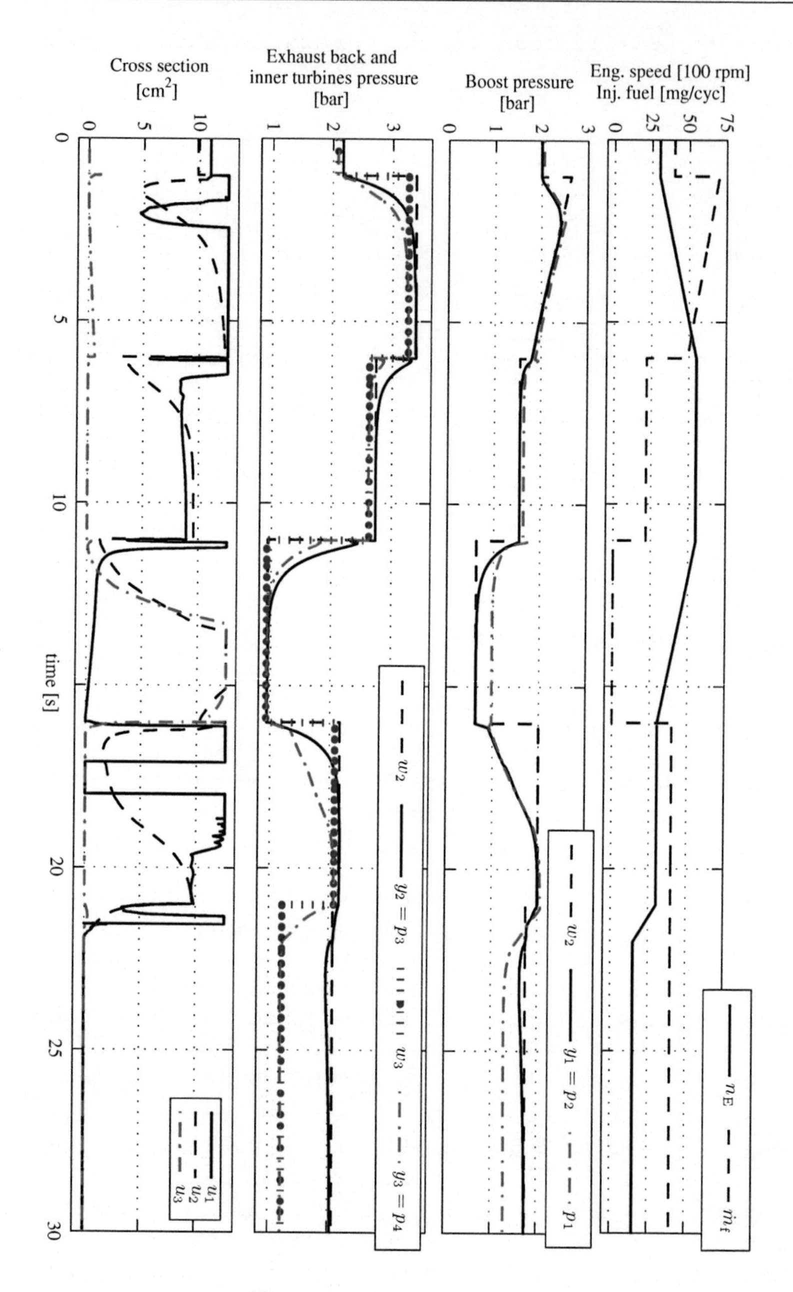

Fig. 7.10: Simulation results

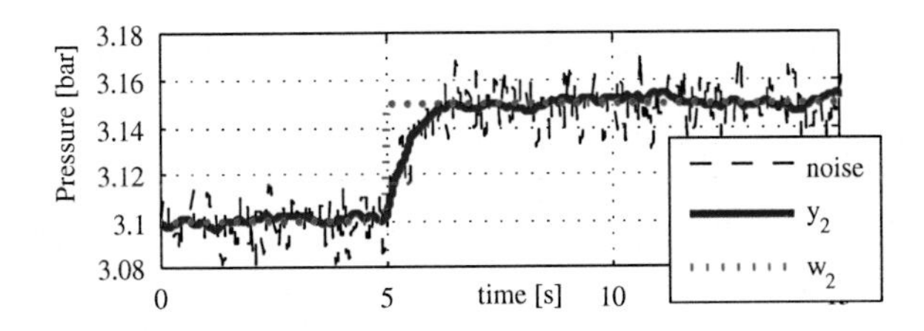

Fig. 7.11: Simulation results of y_2 with $\approx 1\%$ measurement noise

be reproduced by any currently available control solution. At time 16s the air-system is set to attain normal driving conditions for $n_E = 3000$rpm and medium fuel flow. The engine quickly loses speed from time 21s-22s, which resembles a shift in gears. Finally, the engine arrives at 1500rpm in constant driving conditions, for example medium fuel flow and a boost pressure of 2bar. At this point, the HPC delivers almost the entire boost pressure.

When the air-system travels through the singularity at $p_1 = p_2$, the HPC bypass switches between its maximum and minimum values immediately. This happens e. g., at times 16s and 21s. To verify the singularity, the pressure p_1 is also plotted. It can be observed that the HPC bypass opens immediately when the LPC is generating more pressure than the HPC. This is an interesting result, since the theoretically developed IMC controller acts in full accordance to what experts on turbocharging suggest [10].

Figure 7.11 shows the answer of the controlled system with modelling errors under the influence of measurement noise. Noise attenuation and performance can be calibrated online at the engine test bed through the IMC filter poles $-\lambda_i$. As shown here, noise attenuation can be selected to be satisfactory.

In order to show the behaviour of the controlled plant during ageing or production tolerances, some parameters of the plant were altered according to Tab. 7.1. The resulting closed-loop behaviour with various combinations of the parameter variations of Tab. 7.1 is shown in Fig. 7.12 for a random driving scenario. It shows that despite parameter changes, the closed-loop does not become unstable. However, there is a non-zero steady-state offset. This offset comes from the existing input constraints. If the closed-loop IMC controlled two-stage turbocharged air-system does not reach its reference value in steady-state then this means that the

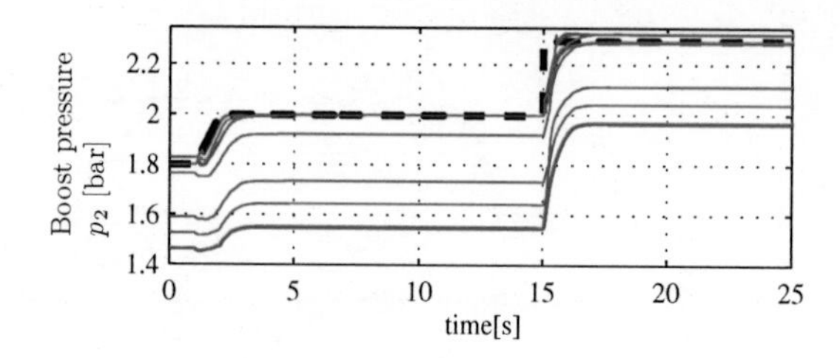

Fig. 7.12: Simulation results of y_2 under some of the plant parameter variations given in Tab. 7.1.

desired reference value is not attainable with the available inputs.

In summary, the results show that the theoretically developed nonlinear IMC control approach successfully solves the difficult control problem. The good tracking performance, the ability of the IMC concept to control the system through its singularities, and to lead it into any physically achievable condition, shows that the nonlinear internal model controller is a very good solution to the control problem of a two-stage turbocharged diesel engine under practical conditions.

7.6 Summary

A nonlinear IMC is developed that controls the boost pressure, the exhaust back pressure, and the pressure between the turbines of a two-stage turbocharged engine. Since the model is input-affine, the model inversion exploited the I/O normal form to develop the IMC controller. However, it is shown, that the inverse becomes singular, if the pressure between the compressors equals the boost pressure. Due to the proposed IMC filter, which was limited to respect the input constraints, the controlled system drives through the singularity. Thus, the setpoints are altered by the IMC filter automatically such that the resulting trajectories for the pressures can be achieved precisely with the available inputs. A stability analysis

Tab. 7.1: Parameter variations

Parameter	Nominal value	Variation
Ambient pressure p_{amb}	0.97 bar	83%-105%
Engine fuel heat value	$1.52 \cdot 10^4 \frac{\text{kJ}}{\text{kg}}$	90%-110%
Exhaust orifice cross section	$9.96\ \text{cm}^2$	70%-100%

showed that the closed-loop is robustly stable. In conclusion, this control solution is a feasible choice as controller of this plant under practical conditions.

8. CONCLUSION

In this thesis, the focus was on the automotive industry and its inherent requirements for controllers. It was established that:

- reference signals are influenced by the driver and, thus, their future values are unknown;
- the closed-loop performance specifications are given by time-domain criteria (as opposed to integral criteria found in optimal control, or demands on the amplitude-over-frequency behaviour of some transfer functions);
- a controller is calibrated (tuned) by non-control engineers at the test bed or in the experimental vehicle and, therefore, needs dedicated tuning-parameters to allow for controller calibration;
- computationally intense operations, such as on-line numerical optimisations, are forbidden due to the weak processing power of OCUs.

There are some control design methods for linear systems which offer the above properties (e. g., loop-shaping, IMC, reference system control design). However, many automotive plants are nonlinear (e. g., catalytic converters, air-systems) and future technologies are expected to introduce more nonlinear plants (e. g., Lean NOx Trap). Therefore, a nonlinear control design method which meets the above requirements is desired.

When it comes to nonlinear controller design, the list of available methods is rather sparse, especially if demanding output feedback control which is a standard requirement on all industrial controllers.

In this thesis, a novel nonlinear control design method was developed by extending the classical IMC design to nonlinear plants. The resulting nonlinear controller fulfills the above requirements and is applicable on stable plants.

The design of internal model control focuses on finding a controller such that a given closed-loop I/O behaviour is achieved. It is based on feedforward control design and relies on model inversion. The IMC structure is simple and plausible and provides valuable properties such as nominal and robust stability as well as zero steady-state offset. The attractiveness of IMC to industry comes from the internal model and the simple design law. The tuning knob of an IMC, as far as nominal performance goes, is reduced to a single parameter that sets the nominal closed-loop bandwidth and results in a well damped control response. Once an IMC controller is determined for a specific plant, it can be calibrated through the model parameters. This enables non-control engineers to calibrate an existing IMC controller, since knowledge of the plant suffices to determine the model parameters. IMC does not use on-line optimisation procedures. Hence, it can be implemented in a real-time environment like an OCU.

The main idea to design a nonlinear IMC controller is to employ the right inverse of the nonlinear plant model together with a low-pass IMC filter. A right inverse is, by itself, a non-realisable operator as it requires a number of derivatives of its input signal. It is the composition of the right inverse with a low-pass filter that yields a non-anticipatory operator which can be used as an IMC controller. The connection between the two is established by implementing the IMC filter as a state-variable filter which gives the requested derivatives by its implementation. It was shown that such an IMC controller yields a nominally stable closed loop which produces zero steady-state offset and has a certain structural robustness.

It was shown that properties of the plant model, like input constraints and singularity of its inverse, can be addressed by the IMC controller if the IMC filter is altered appropriately. To this end, the notion of relative degree of nonlinear systems was exploited. It was possible to describe the shape of the output of a dynamical system in dependence of the shape of the initial conditions, the input function, and the relative degree. With this tool, an algebraic relationship between the input and the highest necessary output derivative was established. This relationship was used to create an IMC filter which, in composition with the right inverse, respects input constraints and does not lead to infinite control responses in the presence of model singularities. Interpreting the IMC filter as a trajectory generator, one finds that the IMC filter is altered such that it is creating only such trajectories that can be a produced by the model with permissible inputs.

A generalisation of this interpretation of the IMC filter was used to propose a novel method to treat models with unstable inverses (non-minimum phase models). The basic idea is founded on the equivalent procedure for linear systems where a zero is used to cancel an unstable pole. The minimum realisation yields an internally stable feedforward controller as the cancellation is done before hand. In the nonlinear case, it was first established under which conditions such a cancellation takes place. This condition was called identical internal dynamics. It was shown that if the IMC filter shares the internal dynamics of the model then the composition and minimal realisation with a perfect right inverse yields a perfectly feasible IMC controller for non-minimum phase nonlinear plants. The conditions on the IMC filter were given and it was suggested how such an IMC filter can be obtained.

With all of the above combined one finds that the proposed nonlinear IMC controller is applicable to a system class which covers many automotive control problems, namely all invertible and stable models, including

- non-minimum phase models,
- models with an ill-defined relative degree,
- models with input constraints
- models with measured disturbances, and
- models with singular inverses.

However, the application of the proposed nonlinear IMC also has some demerits. For good performance, it needs a good control model. With significant modelling errors present, performance is likely to be unsatisfactory (see [22] on the benefit of feedforward control design used together with feedback – an issue related to the IMC design as discussed here). This holds especially for MIMO plants, as dominant cross-couplings (i. e., non diagonally dominant plants) may yield a high sensitivity to modelling errors if inversion based control is employed. IMC may lead to high order controllers. In its minimal realisation, the order of an IMC controller is equal to the order of the model. Clearly, high order plant models automatically lead to a controller of high order. In order to obtain an IMC controller of low order, the order of the plant model needs to be reduced first. Another drawback is, that this thesis has only considered continuous time controllers where, clearly, a discrete time controller needs to be obtained for implementation in an OCU. The task of obtaining a discrete time equivalent is left to the engineer.

Despite these demerits, the presented nonlinear IMC design method is considered very attractive for automotive applications and may be even of interest to other industries.

The nonlinear IMC design method was employed to obtain a controller for two automotive problems. First, the pressure control problem of a one-stage turbocharged diesel engine using a variable-nozzle turbine, and second, the pressure control of three pressures of a two-stage turbocharged diesel engine. Other applications can also be found (see e. g., [viii, xix]).

For the control of the one-stage air-system, the respective model was established to be flat. Therefore, a flatness-based IMC was developed. The IMC filter was chosen to respect input constraints and cancel the effect of measured disturbances. The resulting controller was tested on a real engine at a test bed. The flatness-based IMC compared favourably to the current series production gain-scheduled PID controller in terms of performance and calibration effort. It showed less overshoot and less undershoot in various situations, with comparable closed-loop speed. Hence, it was shown that the proposed control scheme is actually feasible in reality.

For the control of the two-stage turbocharged air-system, only simulation results were available. A nonlinear IMC was developed that controls the boost pressure, the exhaust back pressure, and the pressure between the turbines of a two-stage turbocharged engine. It presents the first control solution to this control problem. Moreover it is the first solution capable to control the plant throughout its physically possible range of operation. Since the model is input-affine, the model inversion exploited the I/O normal form to develop the IMC controller. However, it was shown, that the inverse becomes singular, if the pressure between the compressors equals the boost pressure. Due to the proposed IMC filter, which was limited to respect the input constraints, the controlled system never loses its relative degree. Thus, the setpoints are altered by the IMC filter automatically such that the resulting trajectories for the pressures can be achieved precisely with the available inputs. Moreover, the IMC controller cancelled the effect of measured disturbances. The control performance was very good, even under assumed modelling errors and measurement noise. A stability analysis showed that the closed-loop is robustly stable. It is assumed that the presented controller would perform well under practical conditions as the model quality is similar to that of the one-stage air-system and the controller was obtained using the same control design method.

APPENDIX

A. COMPOSITIONAL MODEL LIBRARY FOR TURBOCHARGED DIESEL ENGINES

This chapter develops a compositional model library for turbocharged air-systems. Models of single turbocharged engines are found, for example, in [14, 52, 70] and follow some different model assumptions and modelling goals as the model presented here. They cannot be used to develop a model of a two-stage turbocharged engine, mainly due to the lack of satisfactory turbine and compressor components. The predominant tools (e. g., [37] or [89]) of modelling engines and air-systems are geared towards component design (as opposed to control system design) and are based on computational fluid dynamics. Computational fluid dynamics describe an air path using finite volumes that are each governed by partial differential equations. The focus is on high-frequency effects like pressure waves and it is not useful for controller design due to the resulting model complexity and since differential equations cannot be extracted from the resulting finite volume model.

The model library, as presented here, is described exhaustively on the next pages. The treatment of the modelling is presented in some detail for the following reasons:

- It is desired to present the reader a self-contained solution to the control problems of a one- and two-stage turbocharged engine. In the opinion of the author, this should include a complete model derivation; especially, since the model library here differs in some aspects from others.

- The library presented here is used to compose the first published model of a two-stage turbocharged air-system that is geared towards control design.

- The turbine and compressor models, as proposed here, are a contribution of this thesis for modelling turbochargers for control applications. Unlike given models in the literature, the presented models are also valid in atypical operating conditions as they appear in two-stage turbocharging.

- Finally, there are no publications on identification and model reduction of turbocharged engines, although those are important steps in obtaining models for control design.

A.1 Modelling Goal and Assumptions

Modelling goal. The goal of modelling an air-system of a turbocharged diesel engine for control design is to obtain

1. a low-frequency model of low order
2. that is composed of re-usable components.

A low-frequency model is important if a model-based controller is to be designed. A compositional approach allows re-use of the component models, to model different turbocharging solutions (e. g., parallel turbines and sequential compressors) simply by re-arrangement.

The modelling focus lies in pressures and temperatures in the pipes and the operation of the turbochargers. The diesel engine is an essential part of this system, however, torque and emission generation are irrelevant considering a pressure control problem and, hence, are not modelled. Similarly, the engine speed n_{E} does not need to be modelled since it is not a control goal and its value is readily available through measurement.

Modelling assumptions. In order to reach the modelling goal, the following assumptions are made:

1. All pipes are assumed to behave like plenum chambers with a uniform distribution of pressure and temperature. This means that high-frequency effects, like pressure waves through a pipe, are neglected and the system can be considered as a lumped parameter system.
2. A mean-value engine model is used, which means that the function of an engine is interpreted as delivering a continuous mass flow, heated by the injected fuel mass. Thus, high-frequency effects, that typically result from a four stroke gas exchange cycle, are neglected.
3. All gases are assumed to be perfect.

Perfect gases follow the law [67]

$$pV = mRT, \tag{A.1a}$$

with pressure p in a chamber of volume V, gas mass m, gas constant R and temperature T. Additionally, specific enthalpy h is the product of constant pressure specific heat c_p and temperature T:

$$h = c_\mathrm{p} T. \tag{A.1b}$$

Finally, the following relationships

$$\kappa = c_\mathrm{p}/c_\mathrm{v} \tag{A.1c}$$

$$R = c_\mathrm{p} - c_\mathrm{v} \tag{A.1d}$$

$$\dot{H} = \dot{m}h, \tag{A.1e}$$

with polytropic exponent κ, constant volume specific heat c_v, total enthalpy flow $\dot{H}$ and mass flow $\dot{m}$ hold.

The following provides the method with which component models are to be connected to one another. More precisely, the physically meaningful input and output signals, through which the components communicate, are defined.

A.2 Connecting Individual Component Models

The component models are divided into two categories, namely

- models with dynamics (storage models), and
- models without dynamics (coupling models).

As a convention, storage models have flow-variables as inputs and some function of their states as outputs. It is not possible to connect energy storage models directly to one another. They can only communicate over a coupling model which provides the matching inputs and outputs. Thus, coupling models compute the flow variables from the outputs of the storage models.

An analysis of Figure 6.1 and Figure 7.1 on pages 125 and 138 shows that there are eight distinct components to be modelled. Those are

- an environment (which is essentially an infinitely large chamber), used to represent V_1, V_5 and V_0, V_6 for a one-stage and a two-stage turbocharged air-system, respectively;

- a plenum chamber, which is used to model the pipes V_2 to V_4 and V_1 to V_5 for a one-stage and a two-stage turbocharged air-system, respectively;
- a turbocharger consisting of turbine, shaft and compressor;
- an intercooler (also referred to as charge air cooler);
- an engine; and
- an orifice, representing bypasses, wastegate and the effect of exhaust gas aftertreatement.

Thus, an air-system can be modelled by few distinct components if they are re-used with different parameters in one model . In the case of the air-system of a turbocharged engine, plenum chambers (the pipes) and the shafts are storage models. Plenum chambers store mass and energy which are represented by the pressure p and temperature T of the fluid inside and are designed to have mass flows $\dot{m}_i$ and enthalpy flows $\dot{H}_i$ of various coupling blocks as inputs. Similarly, shafts have powers P_j as inputs and yield their current speed as output. Figure A.1 shows all storage models used to model the air-systems shown in Figures 6.1 and 7.1. All other components of the air-system are coupling blocks and are

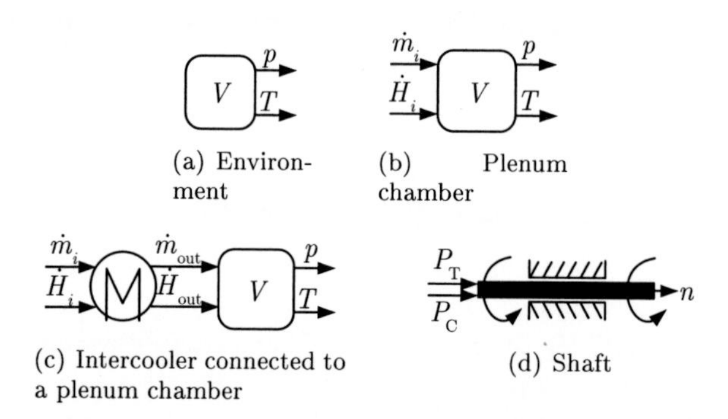

Fig. A.1: Storage component models.

shown in Fig. A.2.

In order to illustrate the connection method, an orifice with constant cross section is placed between two plenum chambers. The resulting connection signals are shown in Figure A.3. It shows the model *structure* of

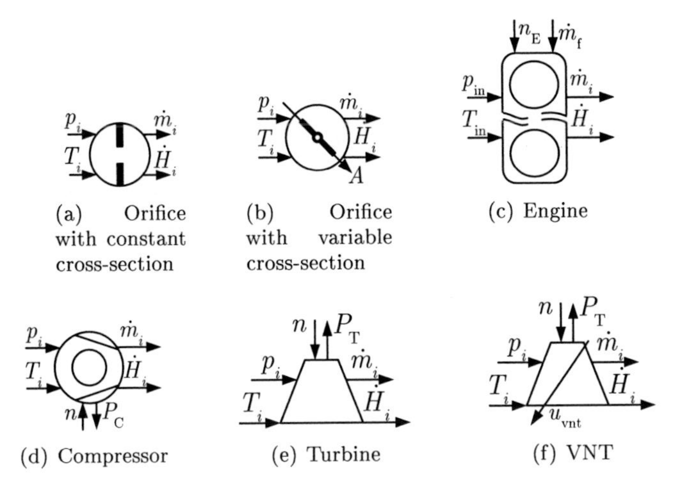

(a) Orifice with constant cross-section (b) Orifice with variable cross-section (c) Engine

(d) Compressor (e) Turbine (f) VNT

Fig. A.2: Coupling component models.

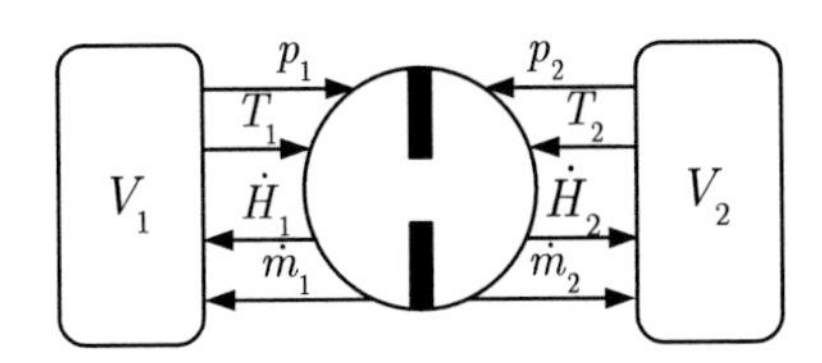

Fig. A.3: Example: Modelling the equalisation between two tanks.

a pressure equalisation procedure between two tanks. Each plenum chamber supplies the orifice with the information of its fluid's current pressure p and temperature T. From this information, the orifice calculates the mass and enthalpy flow from one chamber to the other. The flow information is given to each chamber according to the flow direction: If fluid flows away from a chamber, its flow signals will have a negative sign (e. g., $\dot{m} < 0$ and $\dot{H} < 0$) and if it flows into a chamber the flow signals will be positive.

An exception to the connection rules are intercoolers and environments (see Figure A.1(a)). Intercoolers only lower enthalpy flow and, therefore, have flow-variables as inputs and outputs. They may be regarded as an input-extension of plenum chambers (see Figure A.1(c)). Environments are essentially infinitely large chambers and, therefore, supply a constant pressure and temperature.

A.3 Component Models

A.3.1 Storage Models

Environment

An environment (Fig. A.1(a)) has the constant outputs $p = p_{\text{amb}}$ and $T = T_{\text{amb}}$, representing the ambient pressure p_{amb} and temperature T_{amb}, which are supplied as parameters. Physically, there are always flows into and out of the environment. However, this information is needless from a modelling perspective, since it does not influence the environment.

Plenum chamber with optional intercooler

A derivation of the following equations is based on the energy balance of open systems [67] and the gas law Eq. (A.1a) and can be found e. g., in [92]. Here, a constant volume V is assumed.

Plenum chambers (Fig. A.1(b)) can have an arbitrary number of coupling models connected to them. Each coupling model i sends its flow informations $\dot{m}_i$ and $\dot{H}_i$ to the storage model. The expression $\sum_i$ signifies the sum over all respective flows. The behaviour of a fluid inside a plenum chamber with given parameters V, R, c_{v} and the initial conditions p_0 and T_0 is described by two differential equations:

$$\dot{m} = \sum_i \dot{m}_i, \quad m(0) = \frac{p_0 V}{R T_0} \tag{A.2}$$

$$\dot{p} = \frac{\kappa - 1}{V} \sum_i \dot{H}_i, \quad p(0) = p_0. \tag{A.3}$$

The outputs are given by the pressure p from Eq. (A.3) and the temperature can be obtained from Eq. (A.1a)

$$T = \frac{pV}{mR}, \tag{A.4}$$

with m from Eq. (A.2).

If an intercooler is connected to a chamber (see Figure A.1(c)), it is assumed that it only changes the enthalpy flow or temperature. Thus, the intercooler is assumed to have no pressure drop and does not influence mass flow

$$p_{\text{in}} = p_{\text{out}} \tag{A.5}$$

$$\sum_i \dot{m}_i = \dot{m}_{\text{out}}. \tag{A.6}$$

The coolant has the temperature T_{cool} and lowers the fluid's temperature T relative (by a factor η) to the difference between the two temperatures.

If fluid flows *into* the chamber through the intercooler one finds with Eq. (A.1a)

$$\dot{H}_{\text{out}} = \sum_i \dot{H}_i - \eta \left(\sum_i \dot{H}_i - \dot{Q} \right) \quad \text{with } \dot{Q} = c_{\text{p}} \dot{m} T_{\text{cool}} \text{ for } \sum_i \dot{m}_i > 0. \tag{A.7}$$

However, if fluid flows *out of* a chamber through the intercooler, Eq. (A.7) does not hold. Then, according to Eq. (A.5) the pressure reported from the chamber is correct, but its reported temperature T has to be altered from its actual temperature T_{in}

$$T = T_{\text{in}} - \eta (T_{\text{in}} - T_{\text{cool}}) \text{ for } \sum_i \dot{m}_i < 0 \tag{A.8}$$

and the input signal $\dot{H}_i$ is given to the chamber model unaltered.

Shaft

A shaft (Fig. A.1(d)) has two inputs, namely the power of the turbine P_{T} and the power of compressor P_{C}. With parameters inertia J, damping d, and its initial speed ω_0, the dynamics of a rotating shaft is described by

$$J\dot{\omega} = \frac{1}{\omega} (P_{\text{T}} + P_{\text{C}}) - \omega\, d, \quad \omega(0) = \omega_0 \text{ with } \omega \neq 0, \; \forall t. \tag{A.9}$$

The model output is its current speed ω in radians per second. Since the turbine and compressor models rely on a speed input with unit rpm, the output of the shaft can be chosen as $n = \frac{30}{\pi} \omega$.

A.3.2 Coupling Models

All coupling models receive the information about pressures and temperatures from the plenum chambers they are connected to. They compute mass and enthalpy flows from and to those chambers. Except for the orifice, the coupling models work only in one direction. The variables p_{in} and T_{in} signify the pressure and temperature of the chamber upstream and p_{out} and T_{out} signify the pressure and temperature of the chamber downstream.

Orifice

A model derivation of the flow through an orifice can be found e. g., in [8]. If an orifice is placed between two chambers V_1 and V_2 (cf. Figure A.3) fluid flow is always directed to the lower pressure level p_{out}:

$$\begin{aligned} p_{\mathrm{in}} &= \max\left(p_1, p_2\right) \\ p_{\mathrm{out}} &= \min\left(p_1, p_2\right). \end{aligned}$$

The expressions T_{in} and T_{out} refer to the temperatures in the chambers with the pressures of the same subscript. With the pressure ratio $\Pi = \frac{p_{\mathrm{in}}}{p_{\mathrm{out}}} \geq 1$, the mass flow through the orifice is

$$\begin{aligned} \dot{m} =& A C_{\mathrm{q}} p_{\mathrm{in}} \sqrt{\frac{2}{T_{\mathrm{in}} R}} \cdot \begin{cases} \sqrt{\frac{\kappa}{\kappa-1}\left(\Pi^{-2/\kappa} - \Pi^{\frac{\kappa+1}{-\kappa}}\right)}, & \text{for } \Pi \leq \Pi_{\mathrm{crit}} \\ \left(\frac{2}{\kappa+1}\right)^{\frac{1}{\kappa-1}} \sqrt{\frac{\kappa}{\kappa+1}}, & \text{for } \Pi > \Pi_{\mathrm{crit}} \end{cases} \\ & \text{with } \Pi_{\mathrm{crit}} = \left(\frac{2}{\kappa+1}\right)^{\frac{-\kappa}{\kappa-1}}, \end{aligned} \tag{A.10}$$

where A is the open cross-section area and the parameter C_{q} ($0 \leq C_{\mathrm{q}} \leq 1$) is the flow coefficient which accounts for losses in fluid velocities due to different geometries. The open cross-section A can be variable, representing a continuously variable bypass.

The mass flow signals given to the chambers are

$$\begin{aligned} \dot{m}_{\mathrm{out}} &= -\dot{m} \\ \dot{m}_{\mathrm{in}} &= \dot{m} \end{aligned} \tag{A.11}$$

$$\begin{aligned} \dot{H}_{\mathrm{out}} &= c_{\mathrm{p}} \dot{m}_{\mathrm{out}} T_{\mathrm{out}} \\ \dot{H}_{\mathrm{in}} &= c_{\mathrm{p}} \dot{m}_{\mathrm{in}} T_{\mathrm{out}}, \end{aligned} \tag{A.12}$$

where the subscript "in" and "out" determine which chamber the signal is given to. Equation (A.11) says that mass, which flows from one chamber, must flow into the next. It is further assumed that an orifice is isentropic and, thus, enthalpy flow Eq. (A.12) between the chambers is constant.

Diesel engine

The engine component model (Fig. A.2(c)) essentially covers two effects: It works as a pump and a heater. How much mass is "pumped" through the engine depends upon the air density $\rho_{\mathrm{in}} = \frac{p_{\mathrm{in}}}{RT_{\mathrm{in}}}$ in the intake manifold V_2 and on how much volume can be pushed into the engine at speed n_{E}. With given engine displacement V_{E}, the amount of mass which flows into the engine, on average, during one cycle is

$$m_{\mathrm{E}} = 1/2\,\rho_{\mathrm{in}} V_{\mathrm{E}}(a_0 + a_1 n_{\mathrm{E}}). \tag{A.13}$$

The factor 1/2 in Eq. (A.13) accounts for the fact that air is aspirated only every second cycle. The expression $(a_0 + a_1 n_{\mathrm{E}})$ with parameters a_0 and a_1 accounts for the air-efficiency, which varies in dependence upon the engine speed n_{E}. The linear approximation as presented here is rather rough but has shown to yield sufficient accuracy.

The air mass and enthalpy flow signal given to the intake manifold are

$$\dot{m}_{\mathrm{in}} = -m_{\mathrm{E}} n_{\mathrm{E}} \tag{A.14}$$

$$\dot{H}_{\mathrm{in}} = \dot{m}_{\mathrm{in}} c_{\mathrm{p}} T_{\mathrm{in}}. \tag{A.15}$$

The outgoing mass flow carries the aspirated air and the injected fuel mass $\dot{m}_{\mathrm{f}}$. Combustion does not change mass and, thus, the exhaust mass flow is

$$\dot{m}_{\mathrm{out}} = -\dot{m}_{\mathrm{in}} + \dot{m}_{\mathrm{f}}. \tag{A.16}$$

Combustion heats the outgoing exhaust, however, only a maximum amount of fuel can combust given a certain air mass. With the stoichiometric ratio r_{stoic} and the mass ratio of oxygen r_{O2} in the air the maximum fuel mass flow which can partake in combustion is

$$\dot{m}_{\mathrm{f,maxburn}} = n_{\mathrm{E}} m_{\mathrm{E}} \frac{r_{\mathrm{O2}}}{r_{\mathrm{stoic}}} \tag{A.17}$$

$$\dot{m}_{\mathrm{f,burn}} = \min(\dot{m}_{\mathrm{f}}, \dot{m}_{\mathrm{f,maxburn}}), \tag{A.18}$$

where $\dot{m}_{\mathrm{f,burn}}$ is the actual fuel contributing to combustion. The heat generated by combustion is assumed to be proportional (parameter heat value c) to this fuel mass flow. Thus, the outgoing enthalpy flow is

$$\dot{H}_{\mathrm{out}} = -\dot{H}_{\mathrm{in}} + c\,\dot{m}_{\mathrm{f,burn}} \text{ with } c > 0.$$

Turbine

A turbine (Fig. A.2(e)) cannot be modelled with satisfactory accuracy using only physical or thermodynamic laws, unless a three dimensional computational fluid dynamics approach is chosen. This is especially true for the two-stage turbocharged air-system where the HPT has a large operation range and saturation effects must be taken into account. Therefore, this approach uses a data driven model of a turbine and proposes a new method for extrapolating manufacturer measurement maps.

Turbocharger manufacturers supply measurements of mass flow and efficiency at some operation points of both turbine and compressor. However, this data only represents a small part of the possible operation range and, thus, has to be extrapolated. In [70] several methods for extrapolation are given and compared. Here, a new method of extrapolation is suggested, which yields the necessary accuracy for the two-stage turbocharged diesel engine. It surpasses the models presented in [70] in terms of accuracy, especially concerning the calculation of turbine efficiency.

A turbine's pressure ratio is given as $\Pi = \frac{p_{\text{in}}}{p_{\text{out}}}$, where "in" and "out" refer to the upstream and downstream values, respectively. The isentropic enthalpy flow $\dot{H}_{\text{is}}$ is corrected by the efficiency η to obtain the actual outgoing enthalpy flow $\dot{H}_{\text{out}}$

$$\dot{H}_{\text{is}} = -\dot{H}_{\text{in}} \Pi^{\frac{1-\kappa}{\kappa}} \tag{A.19}$$

$$\dot{H}_{\text{out}} = -\dot{H}_{\text{in}} - \eta \left(\dot{H}_{\text{in}} + \dot{H}_{\text{is}} \right). \tag{A.20}$$

The incoming enthalpy flow $\dot{H}_{\text{in}}$ and the mass flows are defined as

$$\dot{H}_{\text{in}} = c_{\text{p}} T_{\text{in}} \dot{m}_{\text{in}} < 0 \tag{A.21}$$

$$\dot{m}_{\text{in}} = -\dot{m}_{\text{out}} < 0. \tag{A.22}$$

The power P_{T} a turbine supplies to the shaft depends on how much it can lower enthalpy during expansion. One gets

$$P_{\text{T}} = -\dot{H}_{\text{in}} - \dot{H}_{\text{out}} \geq 0. \tag{A.23}$$

With equations (A.19)-(A.23), the ingoing and outgoing signals from the turbine component model are defined if the mass flow $\dot{m}_{\text{out}}$ and efficiency η can be determined.

Extrapolation of mass flow. Manufacturer data for turbines are given in "reduced" mass flow and "corrected" speed, denoted by the subscript "red" and "corr" and are given by

$$\dot{m}_{\text{red}} = \dot{m}_{\text{out}} \frac{\sqrt{T_{\text{in}}}}{p_{\text{in}}} \tag{A.24}$$

$$n_{\text{corr}} = n \sqrt{\frac{T_{\text{ref}}}{T_{\text{in}}}}, \tag{A.25}$$

where T_{ref} is a manufacturer supplied parameter and represents the temperature at which the measurements were recorded. Reduced mass flow $\dot{m}_{\text{red}}$ is given over pressure ratio Π and is assumed to behave similarly as mass flow through an orifice described by Eq. (A.10). The following function approximates $\dot{m}_{\text{red}}$:

$$\dot{m}_{\text{red}} = a(n_{\text{corr}}) \cdot \begin{cases} \sqrt{\frac{\gamma}{\gamma-1}\left(\widetilde{\Pi}^{-2/\gamma} - \widetilde{\Pi}^{\frac{\gamma+1}{-\gamma}}\right)}, & \text{for } \widetilde{\Pi} \le \Pi_{\text{crit}}; \\ \left(\frac{2}{\gamma+1}\right)^{\frac{1}{\gamma-1}} \sqrt{\frac{\gamma}{\gamma+1}}, & \text{for } \widetilde{\Pi} > \Pi_{\text{crit}}. \end{cases} \tag{A.26}$$

$$\text{with } \Pi_{\text{crit}} = \left(\frac{2}{\gamma+1}\right)^{\frac{-\gamma}{\gamma-1}}$$

and with $\widetilde{\Pi} = \Pi - b(n_{\text{corr}})$. The functions $a(n_{\text{corr}})$ and $b(n_{\text{corr}})$ are defined as

$$\begin{aligned} a(n_{\text{corr}}) &= a_1 - a_2 n_{\text{corr}}^2 \\ b(n_{\text{corr}}) &= b_1 n_{\text{corr}}^2. \end{aligned} \tag{A.27}$$

The four parameters γ, a_1, a_2 and b_1 can be determined by fitting the function Eq. (A.26) to measurements. If a variable nozzle turbine (VNT) is used, the four parameters can be assumed to be a function of the nozzle position and can be approximated by splines. Figure A.4 shows an example of an extrapolated turbine mass flow at different speeds n_{corr}.

Extrapolation of efficiency. Efficiency is plotted versus the blade-speed ratio $\Theta = u/c$, where u is the circumferential speed of the rotor and c is the velocity of the fluid flowing out of the turbine

$$\Theta = \frac{u}{c} = \frac{\pi d_{\text{t}}\, n_{\text{red}}/60}{\sqrt{2 c_{\text{p}} T_{\text{ref}} \left(1 - \Pi^{\frac{1-\kappa}{\kappa}}\right)}}. \tag{A.28}$$

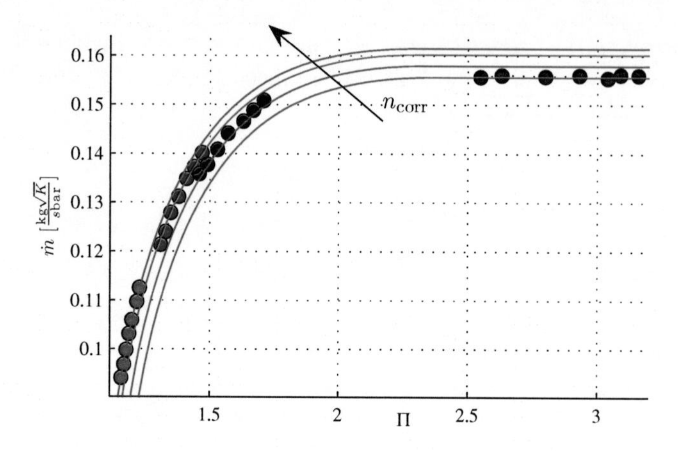

Fig. A.4: Reduced mass flow $\dot{m}_{\text{red}}$ over pressure ratio π at different speeds n_{corr}. The circles show the available data from a manufacturer supplied map and the lines show the behaviour of the model.

The parameter d_{t} signifies the diameter of the turbine's rotor.

In most publications, efficiency η over blade-speed ratio Θ is approximated by a parabola. However, this approach yields unsatisfactory results outside of the provided measurement data since efficiency η must be greater than zero at rotor speed $u = 0 \Leftrightarrow u/c = \Theta = 0$ for $c \geq 0$. Existing turbine models make a fundamental mistake by allowing the efficiency at $\Theta = 0$ to be zero (e.g., [37]) or even negative (e.g., [70]). If this were the case, a turbine would never start rotating when the engine is started. Thus, a novel approximation is presented here. Efficiency is approximated by

$$\eta = \frac{a(n_{\text{corr}})}{b_1 \cdot (\Theta - c(n_{\text{corr}}))^2 + 1} + d_1 \tag{A.29}$$

with

$$\begin{aligned} a(n_{\text{corr}}) &= \frac{a_1}{a_2\,(n_{\text{corr}} - a_3)^2 + 1} \\ c(n_{\text{corr}}) &= c_1 + c_2 n_{\text{corr}}, \end{aligned} \tag{A.30}$$

where the parameters a_1, a_2, b_1, c_1, c_2 and d_1 are determined by fitting the function (A.29) to measurements. If a VNT is used, the six parameters can be assumed to be a function of the nozzle position and can be approximated by splines. Figure A.5 shows an example of an extrapolated turbine efficiency at different speeds n_{corr}.

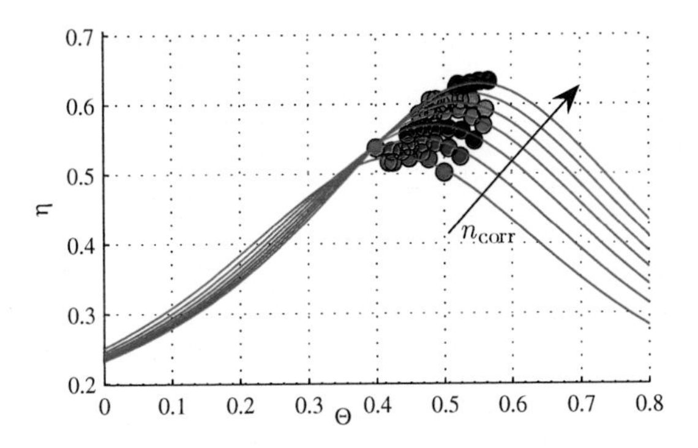

Fig. A.5: Efficiency η over blade-speed ratio Θ at different speeds n_{corr}. The circles show the available data from a manufacturer supplied map and the lines show the behaviour of the model.

Compressor

A compressor's pressure ratio is given as $\Pi = \frac{p_{\mathrm{out}}}{p_{\mathrm{in}}}$. The isentropic enthalpy flow $\dot{H}_{\mathrm{is}}$ is corrected by the efficiency η to obtain the actual outgoing enthalpy flow $\dot{H}_{\mathrm{out}}$

$$\dot{H}_{\mathrm{is}} = -\dot{H}_{\mathrm{in}} \Pi^{\frac{\kappa-1}{\kappa}} \tag{A.31}$$

$$\dot{H}_{\mathrm{out}} = -\dot{H}_{\mathrm{in}} - \frac{1}{\eta}\left(\dot{H}_{\mathrm{in}} + \dot{H}_{\mathrm{is}}\right). \tag{A.32}$$

The incoming enthalpy flow $\dot{H}_{\mathrm{in}}$ and the mass flows are defined as

$$\dot{H}_{\mathrm{in}} = c_{\mathrm{p}} T_{\mathrm{in}} \dot{m}_{\mathrm{in}} < 0 \tag{A.33}$$

$$\dot{m}_{\mathrm{in}} = -\dot{m}_{\mathrm{out}} < 0. \tag{A.34}$$

The power $P_{\mathrm{C}} \leq 0$ a compressor supplies to the shaft depends on how much it can lower enthalpy during expansion. One gets

$$P_{\mathrm{C}} = \dot{H}_{\mathrm{out}} + \dot{H}_{\mathrm{in}} \leq 0. \tag{A.35}$$

With Eqns. (A.31)-(A.35) the ingoing and outgoing signals from the turbine component model are defined if the mass flow $\dot{m}_{\text{out}}$ and efficiency η can be determined.

Extrapolation of mass flow. Compressor mass flow is given "corrected", that is $\dot{m}_{\text{corr}} = \dot{m}\sqrt{\frac{T_{\text{in}}}{T_{\text{ref}}}}\frac{p_{\text{ref}}}{p_{\text{in}}}$. The corrected mass flow is plotted over the pressure ratio Π and is approximated by

$$\dot{m}_{\text{corr}} = a_1 \tan\left(\frac{-\Pi}{b(n_{\text{corr}})}\frac{\pi}{2}\right) + c(n_{\text{corr}}), \tag{A.36}$$

with the functions

$$\begin{aligned} b(n_{\text{corr}}) &= b_1 \tanh(b_2 n_{\text{corr}} + b_3) + b_4 \\ c(n_{\text{corr}}) &= c_1 n_{\text{corr}}^2 + 1, \end{aligned} \tag{A.37}$$

where the parameters a_1, b_1, b_2, b_3, b_4 and c_1 are determined by fitting the function Eq. (A.36) to measurements. Figure A.6 shows an example of an extrapolated compressor mass flow.

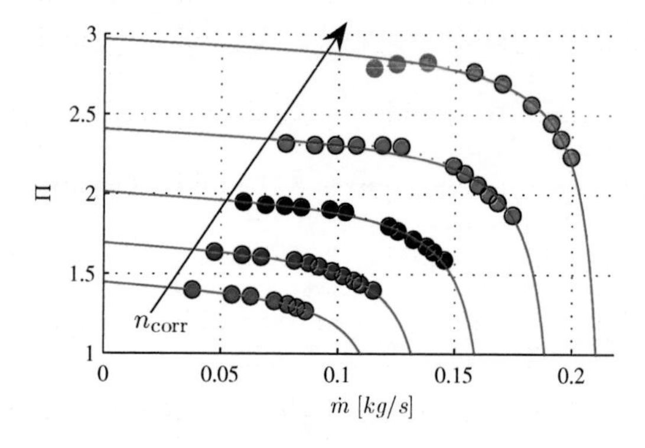

Fig. A.6: Corrected mass flow $\dot{m}_{\text{corr}}$ over pressure ratio Π at different speeds n_{corr}. The circles show the available data from a manufacturer supplied map and the lines show the behaviour of the model.

Extrapolation of efficiency. A compressor's efficiency is plotted over a normalised flow rate Φ with

$$\Phi = \frac{\dot{m}_{\text{corr}}}{\frac{p_{\text{ref}}}{RT_{\text{ref}}}\frac{\pi}{4}d_{\text{c}}^2 u} \text{ with } u = \pi/60 d_{\text{c}} n_{\text{corr}}.$$

The diameter d_c of the compressor rotor is given as a parameter.

Similar to the turbine, most publications suggest a parabola. Again, this approach yields unsatisfactory results. Here, it is proposed to model efficiency by a lower branch of a hyperbola

$$\overline{\eta} = -\frac{\sqrt{a^2 b^2 + a^2(\overline{\Phi} - c)^2}}{b} + d, \tag{A.38}$$

which is rotated by angle α using the substitution

$$\overline{\Phi} \mapsto \cos(\alpha)\Phi - \sin(\alpha)\eta \tag{A.39}$$

$$\overline{\eta} \mapsto \sin(\alpha)\Phi + \cos(\alpha)\eta. \tag{A.40}$$

The parameter a is assumed constant whereas the other variables b, c, d and α are functions of n_{corr}. They can be approximated by splines. Figure A.7 shows an example of an extrapolated compressor efficiency.

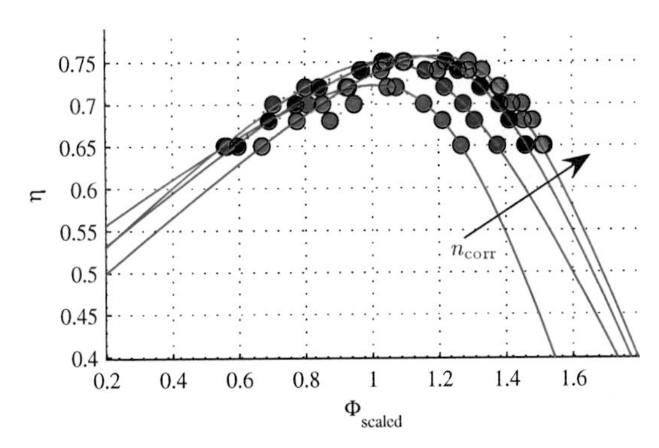

Fig. A.7: Efficiency η over normalised flow rate Φ at different speeds n_{corr}. The circles show the available data from a manufacturer supplied map and the lines show the behaviour of the model.

Figures A.8 and A.9 show the resulting one-stage and two-stage air-system model, respectively, composed using the component models. The arrows indicate the direction of signal flows. Signal flow direction is not necessarily identical to fluid flow direction. Bold arrows indicate vector signals.

Figures A.8 and A.9 present the model *structure* of a one-stage and two-stage turbocharged air-system. Together with the component models introduced in Section A.3 and matching initial conditions, Fig. A.8 and A.9 present simulation models.

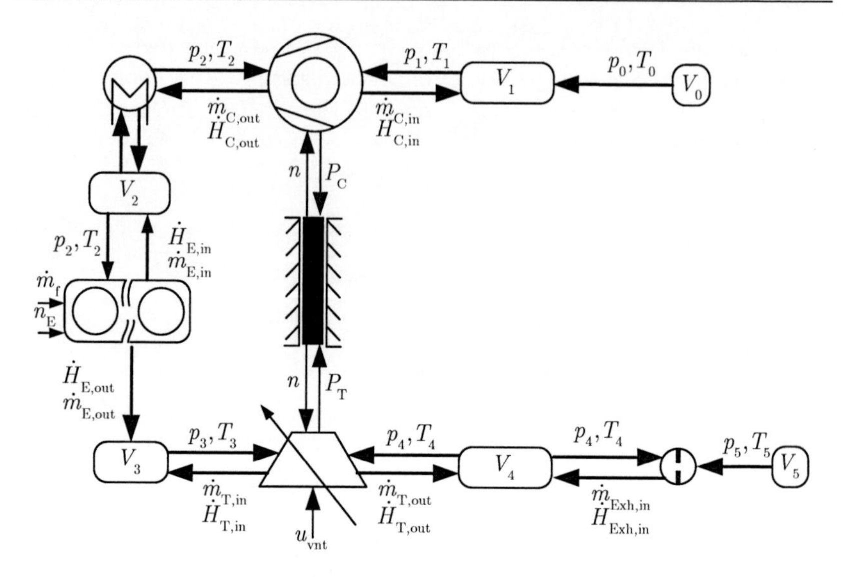

Fig. A.8: Implementation and signal direction of a one-stage turbocharged air-system. Bold arrows indicate vector signals.

A.4 Parameter Estimation and Model Verification

It is desired for the resulting model of an air-system to represent the behaviour of a specific real air-system. Parameter estimation of a continuous-time dynamic model deals with finding the values for some (or all) of the model parameters such that the simulation model (with appropriate initial conditions) represents the behaviour of the real plant. In this work, the criterion used to judge how well the model and real plant fit, is based on the least squares error between the two. Since this work is not concerned with the details of the various optimisation techniques, the reader is referred to [17, 66, 77, 97] and the references therein.

In order to estimate the parameters of the model, at least one measurement of the real system is required. Since such a measurement can be recorded at a test bed, additional sensors can be implemented that should provide data of

- the pressures $p_0, \dots, p_6$, the temperature T_3 and optionally the temperatures $T_0, \dots, T_6$ in all pipes;
- the turbocharger speed(s) n or $n_{\mathrm{LP}}, n_{\mathrm{HP}}$;

- the massflow through the first compressor $\dot{m}_{\mathrm{C,out}}$ or $\dot{m}_{\mathrm{LPC,out}}$;
- and the stimuli of the plant, namely the engine speed n_{E}, fuel massflow $\dot{m}_{\mathrm{f}}$ as well as the chosen VNT position u_{vnt} or the bypass cross sections $A_{\mathrm{HPC}}, A_{\mathrm{HPT}}, A_{\mathrm{LPT}}$.

The following explains the procedure of parameter identification of an air-system. This procedure was found to work well on several engines. On a standard PC, the following algorithm takes a maximum of two hours to complete the parameter estimation for any air-system.

Algorithm A.1 (Parameter estimation of an air-system model [xvi, xvii]**).**

Given: A test bed measurement of the real plant and the model structure shown in Fig. A.8 or Fig. A.9.

Step 1: Identify the engine component model using steady-state signals. *At steady-state conditions, the recorded massflow $\dot{m}_{\mathrm{C,out}}$ or $\dot{m}_{\mathrm{LPC,out}}$ is equal to the fluid flow $-\dot{m}_{\mathrm{E,in}}$ into the engine. The temperature T_3 at steady-state results from engine combustion. Thus, the steady-state measurements suffice to identify all parameters of the engine component model.*

Step 2: Identify the turbine and compressor models. *The turbocharger manufacturer provides the massflow and efficiency maps of the compressor(s) and turbine(s). With these maps, the turbine and compressor component models can be identified. The inertia of a turbocharger is also given in the specifications.*

Step 3: Identify the remaining parameters in the fresh air path. *The model structure can by split into two parts: The fresh air path and the exhaust path. This can be done by regarding the measured turbocharger speed(s) as stimuli to the fresh air path. Therewith, the fresh air path is independent of the exhaust path. It is proposed to estimate the remaining parameters of the fresh air path by using numerical parameter estimation methods. It was found that a Genetic Algorithm [15] works well.*

Step 4: Identify the remaining parameters in the exhaust path. *Similarly to Step 3, with the turbocharger speed(s) as stimuli and the fully identified fresh air path, the exhaust path component models can now be identified using numerical parameter estimation methods.*

Step 5: Identify turbocharger shaft(s) using the full model. *When combining the identified fresh air path with the identified exhaust path, the turbocharger shafts account for the coupling. The parameters of the shaft component model(s) can be estimated using numerical estimation.*

Result: A model of either the one-stage or two-stage turbocharged air-system with a good fit to measurements.

Figure A.10 shows the comparison of simulation and measurement of a one-stage turbocharged engine. Only a small selection of the available signals are shown. However, the portrayed accuracy of the simulation is representative for all signals. The stimuli shown in the first subplot of Fig. A.10 (i. e., engine speed n_{E}, fuel mass flow $\dot{m}_{\mathrm{f}}$ and VNT position u_{vnt}) was chosen such that the whole operation range of the engine was covered. The second subplot of Fig. A.10 shows the comparison of simulated (dashed) and measured (solid) boost pressure p_2. Similarly, subplots three and four portray the massflow through the compressor $\dot{m}_{\mathrm{C,out}}$ (which is equal to the massflow into the system) and the turbocharger speed n, respectively. Overall, the resulting model accuracy is good in both dynamical as well as steady-state behaviour. It is well suited for controller development as it accurately represents the behaviour of the plant at the test bed and is modelled using differential equations.

Figures A.11 to A.13 show the comparison of a two-stage air-system model and the test bed measurements. Again, dashed lines indicate simulated and solid lines indicate measured signals. The available test bed measurement from the two-stage turbocharged diesel engine lacks some important data. First, the engine is not stimulated in its whole operating range. Second, important pressure signals, namely p_5 and p_6 were not available. Third, the bypass of the HPC was not closing correctly and the bypass cross sections were measured poorly. Unfortunately, better measurement data is not available for this work. Despite the bad quality of the available measurement data, the resulting model is chosen for controller development.

The computational effort for simulating the models is low: On a 3Ghz Pentium, the one-stage simulation model needs below one second computation time for about 500 seconds of simulation time. The two-stage model needs about three seconds for the same simulation time interval. Therefore, one concludes that it is possible to simulate both models in real-time on a controller that is about 100 times slower than a 3Ghz Pentium.

A.5 Model Simplification

This section is concerned with reducing the order of the air-system models. Model reduction is an important step, because a model of low order greatly simplifies nonlinear IMC design and reduces the order of the resulting IMC controller.

The order of a coupled air-system model results from the number of chamber component models (two states per chamber, see equations (A.2) and (A.3)) and the number of shafts (one state per shaft, see Eq. (A.9)). This yields a *ninth order* model for the one-stage turbocharged air-system (cf. Fig. A.8) and a *twelfth order* model for the two-stage turbocharged air-system (cf. Fig. A.9).

Considering the dynamics of the storage component models with realistic parameterisation, it becomes clear that the chamber models have significantly faster dynamics than the shaft models: Tests using the simulation models have shown that the volume of any chamber model has virtually no effect on the dynamics of the simulation result in the considered low-frequency domain. This holds for chamber volumes lower than about fifty liters, where fifty liters is an unrealistically high number for the volume of a chamber in any air-system. In conclusion, the air-system low-frequency dynamics solely stem from the dynamics of the shaft component models. Since this work is only interested in the low-frequency behaviour of the air-system, it is proposed to reduce the model order by replacing either one or both states of the chamber component models with algebraic relationships.

A.5.1 Simplified Chamber Model

Steady-state temperature dynamics in the plenum chamber

During driving, fluid is constantly flowing through the pipes in the same direction. This simplification assumes that the outgoing fluid's temperature is equal to the incoming fluid's temperature. Thus, the simplified chamber component model is described by

$$T = \frac{\sum_{\text{in}} \dot{H}_{i,\text{in}}}{c_{\text{p}} \cdot \sum_{\text{in}} \dot{m}_{i,\text{in}}} \tag{A.41}$$

$$\dot{p} = \frac{\kappa - 1}{V} \sum_i \dot{H}_i, \quad p(0) = p_0, \tag{A.42}$$

where the subscript "in" indicates that only incoming fluid flow is considered. Thus, the terms $\sum_{\text{in}} \dot{H}_{i,\text{in}}$ and $\sum_{\text{in}} \dot{m}_{i,\text{in}}$ mean the sum over all *positive* enthalpy flows $\dot{H}_i$ (with $\dot{H}_i > 0$) and *positive* mass flows $\dot{m}_i$ (with $\dot{m}_i > 0$), respectively. By replacing the original chamber component model, described by Eqns. (A.2)-(A.4) with Eqns. (A.41) and (A.42), each plenum chamber will consist of only one state, significantly reducing the number of states in the implementation of an air-system model.

Steady-state model of a plenum chamber

This simplification aims at replacing the pressure dynamics (A.42) by an algebraic relationship. Together with the simplified temperature calculation (A.41) this step results in a completely algebraic model of a plenum chamber [xvi].

It is assumed that temperature and pressure in a chamber are directly related to each other by the polytropic relationship

$$p = \left(\frac{T}{c}\right)^{\frac{k}{k-1}} \tag{A.43}$$

where c and k are parameters which vary for each chamber and have to be determined separately for each chamber.

The assumption in Eq. (A.43) only holds in the fresh air path but not in the exhaust path. This is due to the effect of fuel combustion on the pressure in a chamber in the exhaust path, for which no good algebraic approximation was found. In the fresh air path, this simplification yields a good approximation of the behaviour of a chamber.

A.5.2 Model Simplification Procedure

The following develops the simplification of a two-stage turbocharged air-system and applies to Fig. A.9. With only minor changes, the steps introduced below are also applicable for the one-stage turbocharged air-system given in Fig. A.8.

Simplification of the fresh air path

The fresh air path does not model the turbocharger dynamics and rather assumes the turbocharger speeds as given exogenous signals. It follows that the chambers are the only dynamic components in the fresh air path and that their dynamics can be neglected, one concludes that the complete fresh air path from environment up to and including the engine could essentially be represented algebraically. However, such an approach yields algebraic loops which have to be solved numerically. Therefore, the following develops a simplification of the fresh air path which keeps the boost pressure p_2 as a state and does not yield an algebraic loop:

1. The boost pressure p_2 and boost temperature T_2 are calculated by the equations (A.42) and (A.43), which means that p_2 remains a state. This gives the signals p_2 and T_2.

2. From Eq. (A.1b), the density in the chamber V_2 is available by $\rho_2 = p_2/(RT_2)$ and the engine model can compute the fluid and enthalpy flows $\dot{m}_{\mathrm{E,out}}$ or $\dot{H}_{\mathrm{E,out}}$, respectively.

3. The fluid flow through the engine $\dot{m}_{\mathrm{E,out}}$, without the injected fuel mass $\dot{m}_{\mathrm{f}}$ is assumed to flow through the LPC:

$$\dot{m}_{\mathrm{LPC,out}} \approx \dot{m}_{\mathrm{E,out}} - \dot{m}_{\mathrm{f}}$$

   With the output $\dot{m}_{\mathrm{LPC,out}}$ of the compressor model from the equation above, the pressure ratio Π of the compressor can be computed from Eq. (A.36). With the known ambient pressure p_0, the pressure p_1 results from $p_1 = \Pi p_0$. Now, all input signals of the LPC are known and its output signal $\dot{H}_{\mathrm{LPC,out}}$ can be computed.

4. The temperature T_1 follows using Eq. (A.41).

5. All input signals for the HPC are known and its remaining output signals $\dot{m}_{\mathrm{HPC,out}}$ and $\dot{H}_{\mathrm{HPC,out}}$ follow from the component model.

6. Finally, all inputs signals for the HPC bypass are known and its outputs $\dot{m}_{\mathrm{OHPC,out}}$ and $\dot{H}_{\mathrm{OHPC,out}}$ follow from the component model.

This results in a fresh air path, consisting of a single state (namely p_2) instead of four states. Moreover, no algebraic loop is present.

Simplification of the exhaust path

In the exhaust path, a similar reduction of the states is not possible, since Eq. (A.43) does not hold. Thus, in order to obtain a reduced model without any algebraic loops, more states will have to be retained.

1. The temperature T_3 in the exhaust manifold follows from Eq. (A.41) using the known output signals $\dot{m}_{\mathrm{E,out}}$, $\dot{H}_{\mathrm{E,out}}$ from the engine. The exhaust back pressure p_3 is kept as a state and computed according to Eq. (A.42). Therewith, the signals p_3 and T_3 are given.

2. In order to obtain an input-affine model of a two-stage turbocharged engine, the chamber V_4 is not simplified. Thus, both p_4 and T_4 remain states.

3. Experience shows that the temperature T_5 varies only slightly. Thus, T_5 is assumed to be constant. Finally, with the assumptions that the pressure in the exhaust pipe p_5 is higher than the ambient pressure $p_5 > p_6$ and that the flow out of the exhaust pipe is approximately equal to the massflow out of the engine

$$-\dot{m}_{\mathrm{Exh,in}} \approx \dot{m}_{\mathrm{E,out}},$$

   the pressure p_5 can now be computed from Eq. (A.10).

As a result, the simplified exhaust path consists of three states (p_3, p_4, T_4) instead of the original six.

The coupling of the simplified fresh air path with the simplified exhaust path is done using the shaft component models which are not changed. In summary, the simplified model of a two-stage turbocharged air-system consists of six states, namely $p_2, p_3, p_4, T_4, \omega_{\mathrm{LP}}$, and ω_{HP}. Thus, six states of the original model have been removed.

A.6 Control Design Models of One- and Two-Stage Turbocharged Diesel Engines

A.6.1 Simplified Model of a One-Stage Turbocharged Diesel Engine

A similar simplification as the one developed for the two-stage turbocharged air-system can be employed to simplify the one-stage turbocharged air-system (cf. Fig. A.8). An extensive review of the simplification can be found in [76] and will not be treated here in detail.

This simplified model is used as IMC model $\widetilde{\Sigma}$ and is the basis for developing a feedforward controller. It is described by using the state vector $\boldsymbol{x} = [\omega, p_2]^T$ and the input $u = u_{\mathrm{vnt}}$, where ω is the speed of the turbocharger and p_2 is the boost pressure. The input u is a function of the nozzle position of the variable nozzle turbine (VNT). The engine speed n_E and the injected fuel mass $\dot{m}_\mathrm{f}$ are considered to be measured disturbances $\boldsymbol{d}_\mathrm{m} = [n_\mathrm{E}, \dot{m}_\mathrm{f}]^T$. The model $\widetilde{\Sigma}$ is given by

$$\begin{aligned}
\widetilde{\Sigma}: \ \dot{x}_1 =& \frac{k_1 k_2}{x_1}\left(k_3(\boldsymbol{d}_\mathrm{m}) + k_6(\boldsymbol{d}_\mathrm{m}) k_4 x_2\right)\varphi_1(x_2) - \frac{k_1 k_5 \varphi_3(x_2)}{x_1}\varphi_2(x_1, x_2) \\
&- \frac{k_1 k_2}{x_1}\left(k_3(\boldsymbol{d}_\mathrm{m}) + k_6(\boldsymbol{d}_\mathrm{m}) k_4 x_2\right)\varphi_1(x_2) u \\
\dot{x}_2 =& \frac{\varphi_4(x_2)}{k_7}\varphi_2(x_1, x_2) - \frac{k_9}{k_7} k_6 x_2
\end{aligned} \tag{A.44a}$$

with the initial condition $\boldsymbol{x}(0) = \boldsymbol{x}_0$ and the output equation

$$\tilde{y} = x_2 = p_2 \tag{A.44b}$$

with

$$\begin{aligned}
\varphi_1(x_2) &= \frac{k_9}{k_4} + \frac{k_{17}(\boldsymbol{d}_\mathrm{m})}{k_{14}(\boldsymbol{d}_\mathrm{m}) + k_4 x_2} \triangleq T_3 \\
\varphi_2(x_1, x_2) &= k_8 \frac{k_{10} x_1^2 - \varphi_3(x_2)}{k_{11} x_1} \triangleq \dot{m}_\mathrm{C,out} \\
\varphi_3(x_2) &= k_{16}\left(\left(\frac{x_2}{p_\mathrm{amb}}\right)^{k_{12}} - 1\right) \triangleq \frac{\dot{H}_\mathrm{C,out} + \dot{H}_\mathrm{C,in}}{\dot{m}_\mathrm{C,out}} \\
\varphi_4(x_2) &= k_{15}\varphi_3(x_2) + k_{13} \triangleq T_2,
\end{aligned}$$

where the coefficients k_i are system parameters (e. g., diameters, inertia of the turbocharger, etc.), p_amb is the ambient pressure and the $\varphi_j(\cdot)$ are some nonlinear functions of the states.

Figure A.14 shows the simplified model versus the original model and measurement data. Although the sixth order model was reduced to second order, the accuracy is still good. In conclusion, the reduced-order model (6.1) can be used for controller design.

A.6.2 Simplified Model of a Two-Stage Turbocharged Diesel Engine

The six state variables of the reduced model of a two-stage turbocharged engine are explained in Tab. A.1 and Tab. A.2 introduces the parameters.

Tab. A.1: State variables of the simplified plant model of the two-stage turbocharged engine (cf. Fig. A.9).

x_1:	Boost pressure p_2
x_2:	Exhaust back pressure p_3
x_3:	Pressure between turbines p_4
x_4:	Fluid mass m_4 contained in pipe V_4
x_5:	Speed ω_{LP} of LP shaft
x_6:	Speed ω_{HP} of HP shaft

Tab. A.2: Nomenclature for the two-stage turbocharged air-system (cf. Fig. A.9).

$\boldsymbol{d}_{\mathrm{m}}$:	Measured disturbances, engine speed and fuel mass flow ($\boldsymbol{d}_{\mathrm{m}} = [n_{\mathrm{E}}, \dot{m}_{\mathrm{f}}]^T$)
κ_i:	Chamber parameter polytropic exponent of chamber i
V_i:	Chamber parameter volume of chamber i
$J_{\mathrm{LP}}, J_{\mathrm{HP}}$:	Shaft parameter inertia of low-pressure or high-pressure shaft, respectively
$d_{\mathrm{LP}}, d_{\mathrm{HP}}$:	Shaft parameter damping of low-pressure or high-pressure shaft, respectively

The reduced order, input-affine model $\widetilde{\Sigma}$, with which the IMC controller is developed, is given by

$$\widetilde{\Sigma}: \quad \dot{\boldsymbol{x}} = \boldsymbol{f}(\boldsymbol{x}, \boldsymbol{d}_{\mathrm{m}}) + \boldsymbol{G}(\boldsymbol{x}, \boldsymbol{d}_{\mathrm{m}})\boldsymbol{u}, \;\; \boldsymbol{x}(0) = \boldsymbol{x}_0, \;\; \boldsymbol{x} \in \mathbb{R}^6 \; \boldsymbol{u} \in \mathbb{R}^3 \tag{A.45a}$$

$$\tilde{\boldsymbol{y}} = \boldsymbol{h}(\boldsymbol{x}) = \begin{bmatrix} x_1, & x_2, & x_3 \end{bmatrix}^T, \;\; \boldsymbol{y} \in \mathbb{R}^3, \tag{A.45b}$$

with model output $\tilde{\boldsymbol{y}}$. The vector field $\boldsymbol{f}$ is defined as

$$\boldsymbol{f}(\boldsymbol{x}, \boldsymbol{d}_{\mathrm{m}}) = \begin{bmatrix} \frac{\kappa_2 - 1}{V_2}\left(\dot{H}_{\mathrm{HPC,out}} + \dot{H}_{\mathrm{E,in}}\right) \\ \frac{\kappa_3 - 1}{V_3}\left(\dot{H}_{\mathrm{E,out}} + \dot{H}_{\mathrm{HPT,in}}\right) \\ \frac{\kappa_4 - 1}{V_4}\left(\dot{H}_{\mathrm{HPT,out}} + \dot{H}_{\mathrm{LPT,in}}\right) \\ \dot{m}_{\mathrm{HPT,out}} + \dot{m}_{\mathrm{LPT,in}} \\ \frac{1}{J_{\mathrm{LP}}\omega_{\mathrm{LP}}}\left(P_{\mathrm{LPT}} - P_{\mathrm{LPC}} - d_{\mathrm{LP}}\omega_{\mathrm{LP}}^2\right) \\ \frac{1}{J_{\mathrm{HP}}\omega_{\mathrm{HP}}}\left(P_{\mathrm{HPT}} - P_{\mathrm{HPC}} - d_{\mathrm{HP}}\omega_{\mathrm{HP}}^2\right) \end{bmatrix} \tag{A.45c}$$

and the matrix $\boldsymbol{G}$ has the following structure:

$$\boldsymbol{G}(\boldsymbol{x},\boldsymbol{d}_{\mathrm{m}})=\begin{bmatrix} g_{11}(\boldsymbol{x},\boldsymbol{d}_{\mathrm{m}}) & 0 & 0 \\ 0 & g_{22}(\boldsymbol{x},\boldsymbol{d}_{\mathrm{m}}) & 0 \\ 0 & g_{32}(\boldsymbol{x},\boldsymbol{d}_{\mathrm{m}}) & g_{33}(\boldsymbol{x},\boldsymbol{d}_{\mathrm{m}}) \\ 0 & g_{42}(\boldsymbol{x},\boldsymbol{d}_{\mathrm{m}}) & g_{43}(\boldsymbol{x},\boldsymbol{d}_{\mathrm{m}}) \\ 0 & 0 & 0 \\ 0 & 0 & 0 \end{bmatrix} \cdot 1/\mathrm{m}^2 \tag{A.45d}$$

with

$$\begin{aligned} g_{11} &= \frac{\kappa_2 - 1}{V_2}\dot{H}^*_{\mathrm{OHPC,out}} \\ g_{22} &= \frac{\kappa_3 - 1}{V_3}\dot{H}^*_{\mathrm{OHPT,in}} \\ g_{32} &= \frac{\kappa_4 - 1}{V_4}\dot{H}^*_{\mathrm{OHPT,out}} \\ g_{33} &= \frac{\kappa_4 - 1}{V_4}\dot{H}^*_{\mathrm{OLPT,in}} \\ g_{42} &= \dot{m}^*_{\mathrm{OHPT,out}} \\ g_{43} &= \dot{m}^*_{\mathrm{OLPT,in}}, \end{aligned} \tag{A.45e}$$

where the asterisk '$*$' indicates that the respective flow is to be calculated assuming a unitary (i. e., $1\mathrm{m}^2$) open cross section of the respective bypass. For example, $\dot{H}^*_{\mathrm{OHPC,out}}$ presents the output enthalpy flow of the high-pressure compressor bypass with a cross section of $1\mathrm{m}^2$. From $\dot{H}_{\mathrm{OHPC,out}} = \dot{H}^*_{\mathrm{OHPC,out}} \cdot 1/\mathrm{m}^2 \cdot u_1$ one finds that the unit correction $1/\mathrm{m}^2$ is necessary since the input u_1 has units of m^2.

All enthalpy flows $\dot{H}_{(\cdot)}$ and mass flows $\dot{m}_{(\cdot)}$ in Eqns. (7.4c)-(A.45e) are given from the coupling component models as introduced in Section A.3. Thus, they are functions dependent upon the states of their neighbouring storage component models and upon the measured disturbances $\boldsymbol{d}_{\mathrm{m}} = [n_{\mathrm{E}}, \dot{m}_{\mathrm{f}}]^T$.

This leads to an interesting result: The model structure shown in Fig. A.9 and the model equations (7.4) are both independent of the specifics of the coupling or storage component models. Thus, different models of turbines, compressors, orifices, chambers do not change the model structure as long as no additional states are introduced.

The quality of the reduced-order model is shown in Fig. A.15 by comparing simulation results of the full and reduced-order model. The reduced-order model is such a good approximation that e. g., the boost pressure never deviates more than 0.02 bar from the full order model. In conclusion, the model (7.4) is feasible for controller development.

A.7 Summary

This chapter has introduced a compositional model library for turbocharged air-systems which is able to represent their low-frequency behaviour well. With the library and the presented algorithms of parameter identification and model reduction, it is possible to obtain coupled models for virtually any turbocharging solution and use the resulting model for controller design.

The library components are divided into storage and coupling models where only storage models have a dynamic behaviour. Thus, storage models are represented by differential equations and coupling models are represented by algebraic equations. The connection philosophy allows to build air-systems by alternatingly connecting storage models to coupling models. Therefore, an algorithm has been introduced which simplifies parameter estimation of a model, given an appropriate test bed measurement and the turbocharger maps. The test bed measurement must contain all pressures, the temperature in the exhaust manifold, the turbocharger speeds and the stimulation signals of engine speed, fuel mass flow and the VNT position or bypass cross section signals. The main idea of the parameter identification is to split the air-system into the fresh air path and the exhaust path by using the measured turbocharger speeds as stimuli. Therewith, the fresh air path and the exhaust path can be identified separately, reducing the number of parameters to be identified at once.

The resulting model quality of both a one-stage and a two-stage turbocharged air-system is very good. However, the models are of ninth and twelfth order for the one- and two-stage model, respectively. It is proposed to reduce the model order by using simplified component models for the plenum chambers since their dynamics can be neglected. An algorithm for model reduction is given and the resulting models show almost no deviation from the models they originated from. Finally, it is proposed to use the simplified models for a design of an internal model controller.

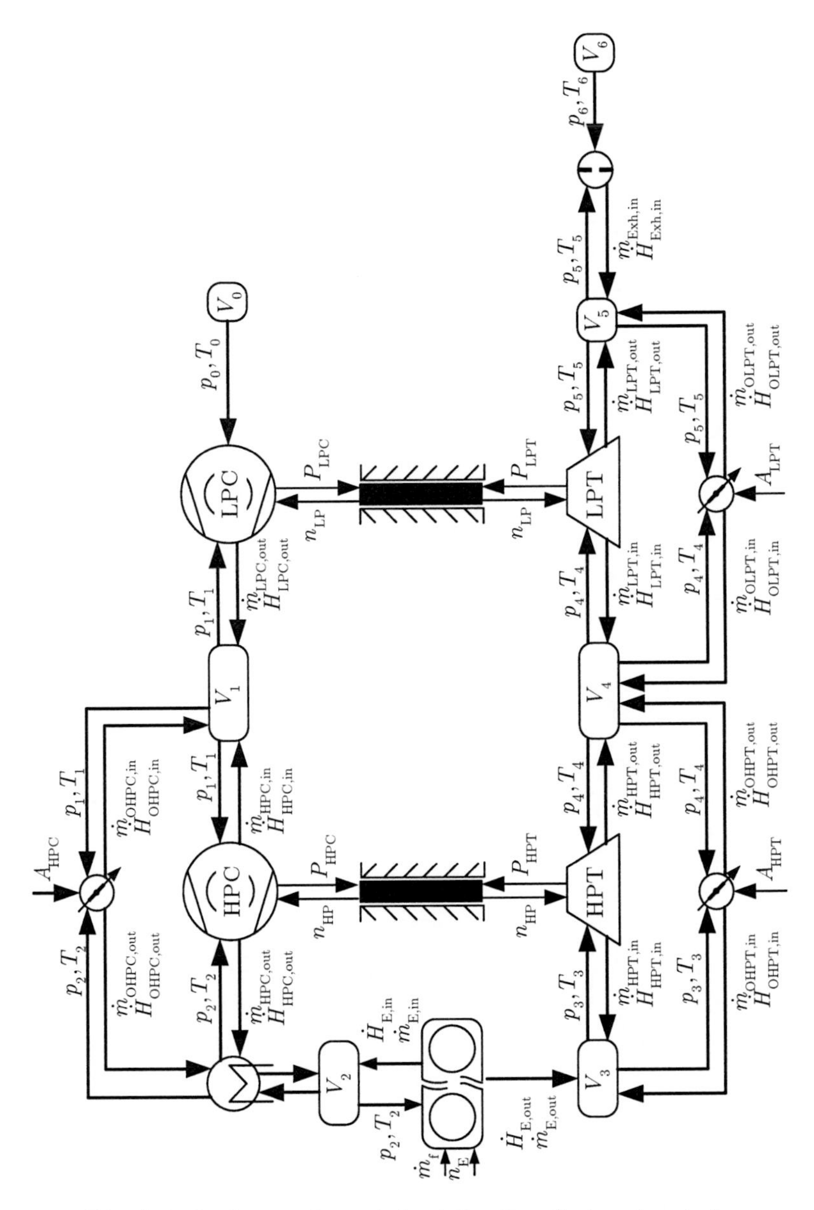

Fig. A.9: Implementation and signal direction of a two-stage turbocharged air-system. Bold arrows indicate vector signals.

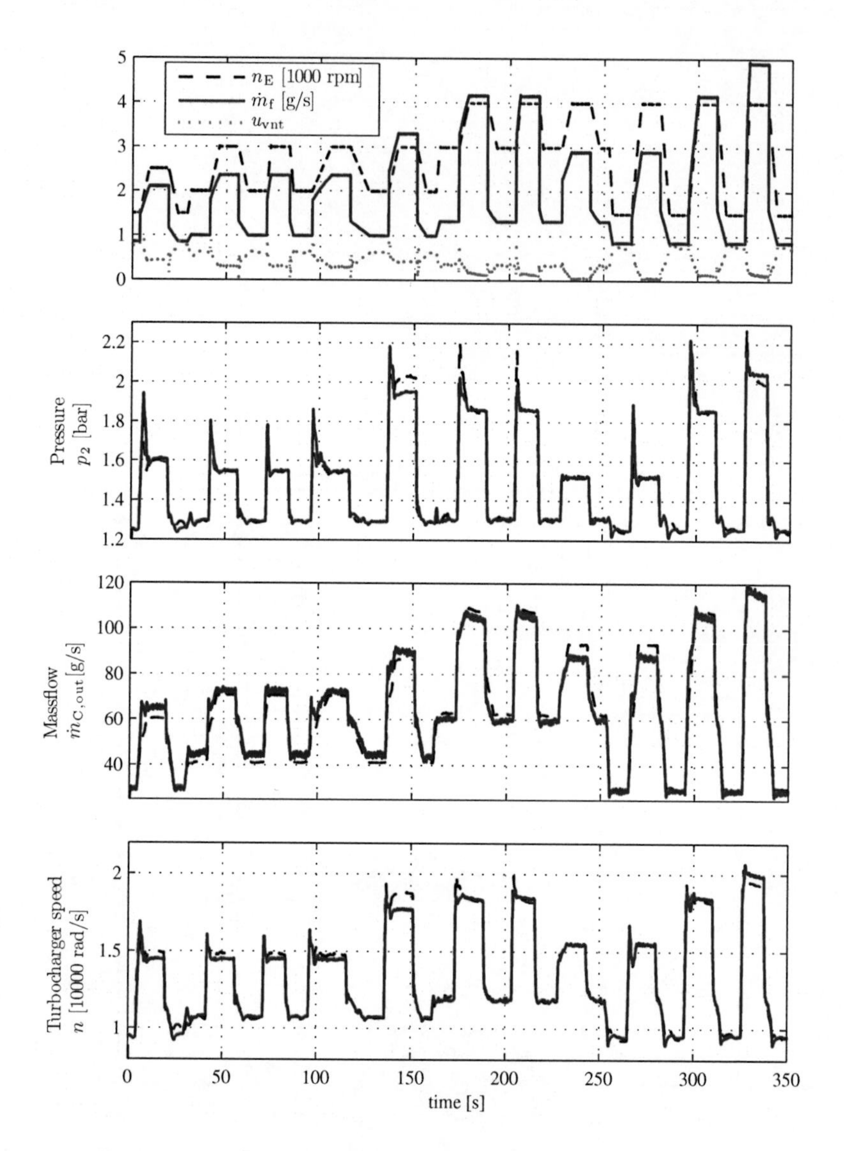

Fig. A.10: Model validation of a one-stage turbocharged diesel engine. Dashed lines indicate simulation results and solid lines refer to measurement.

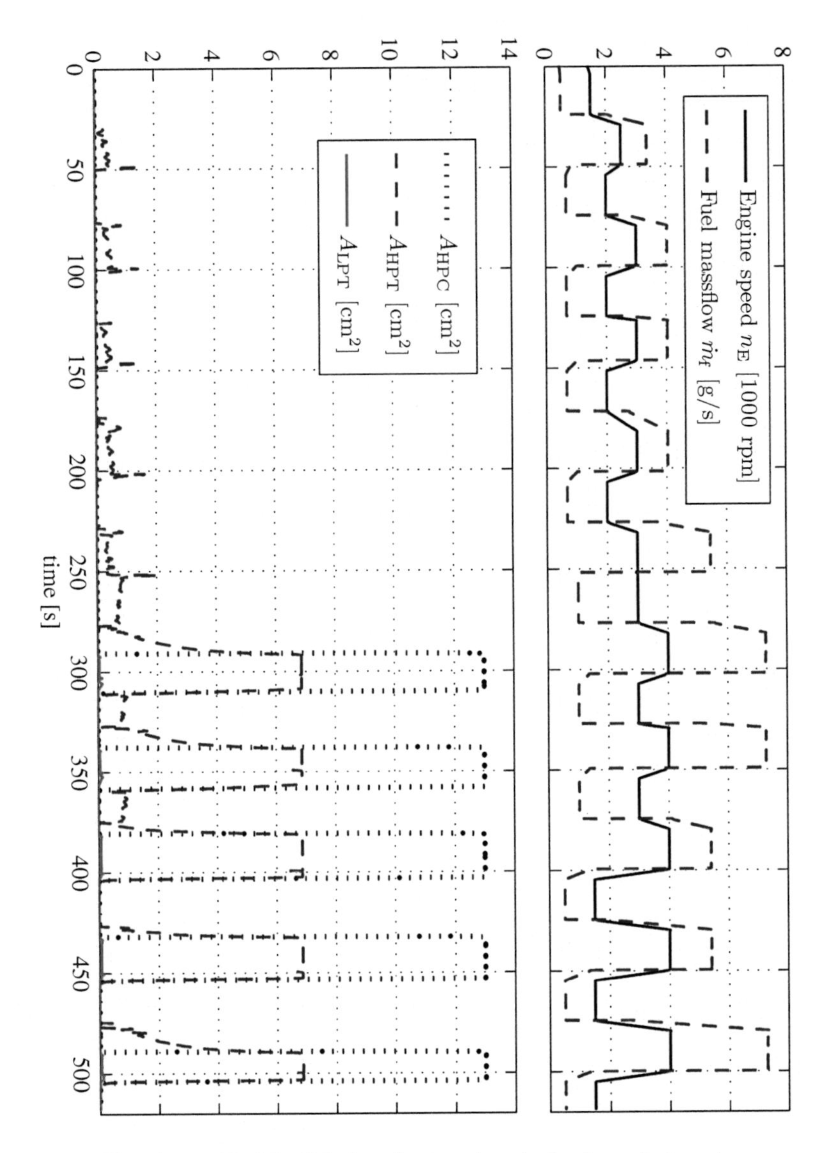

Fig. A.11: Model validation of a two-stage turbocharged air-system (1 of 3). Engine speed n_E and fuel mass flow $\dot{m}_f$ are shown.

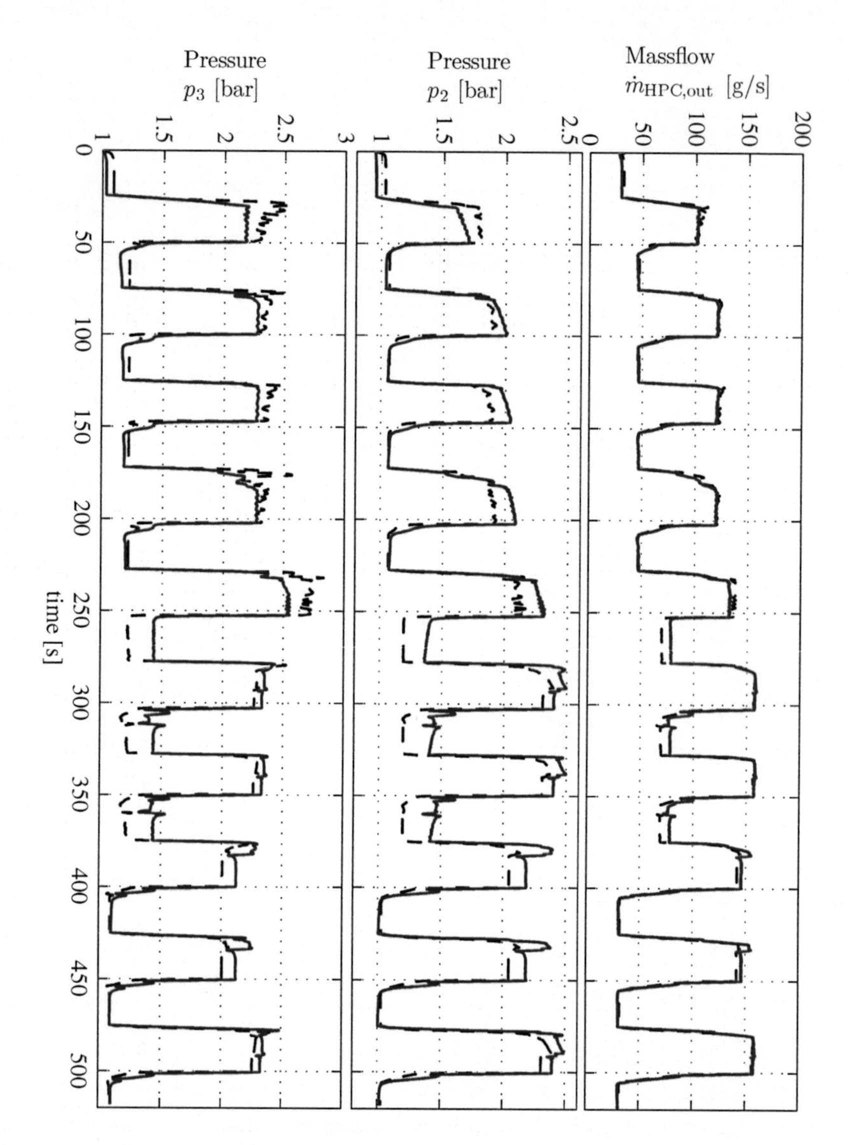

Fig. A.12: Model validation of a two-stage turbocharged air-system (2 of 3). The massflow through the LPC $\dot{m}_{\mathrm{LPC,out}}$, the boost pressure p_2 and the exhaust back pressure p_3 are shown. Dashed lines indicate simulation results and solid lines refer to measurement.

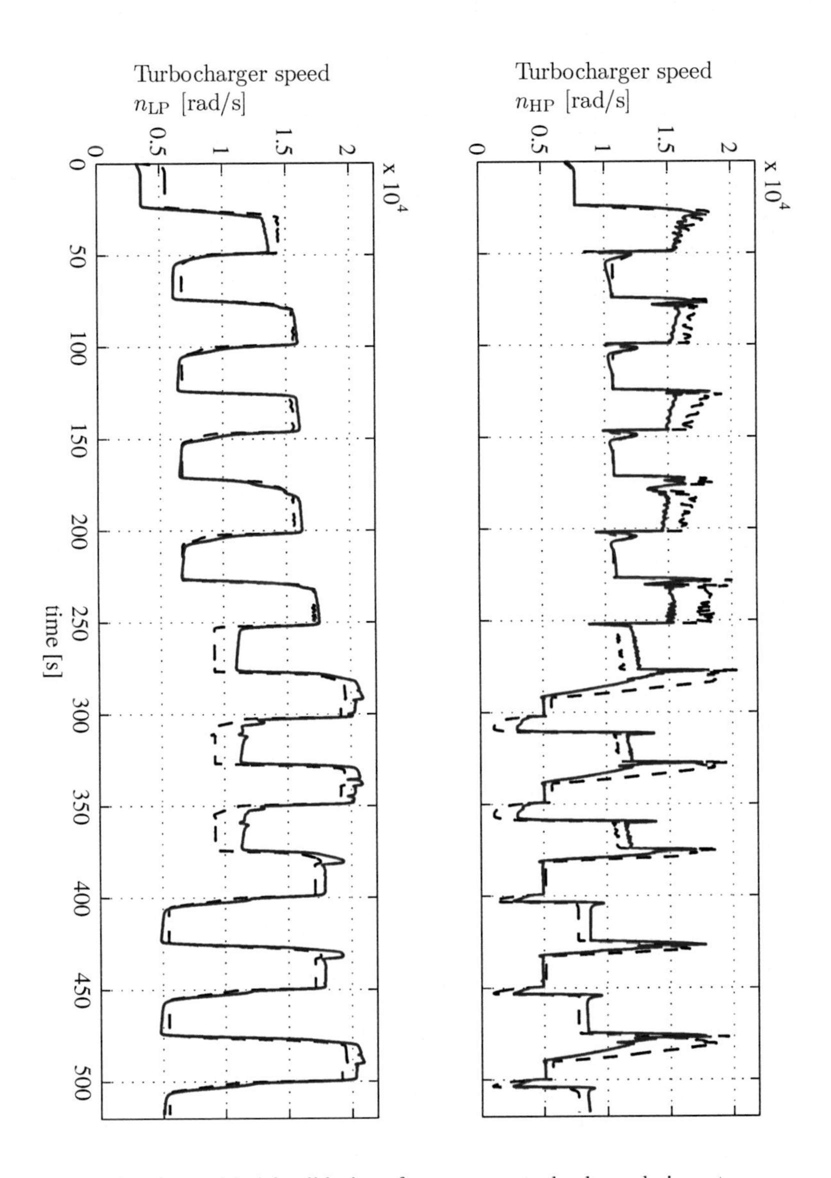

Fig. A.13: Model validation of a two-stage turbocharged air-system (3 of 3). The turbocharger speeds n_{HP} and n_{LP} are shown. Dashed lines indicate simulation results and solid lines refer to measurement.

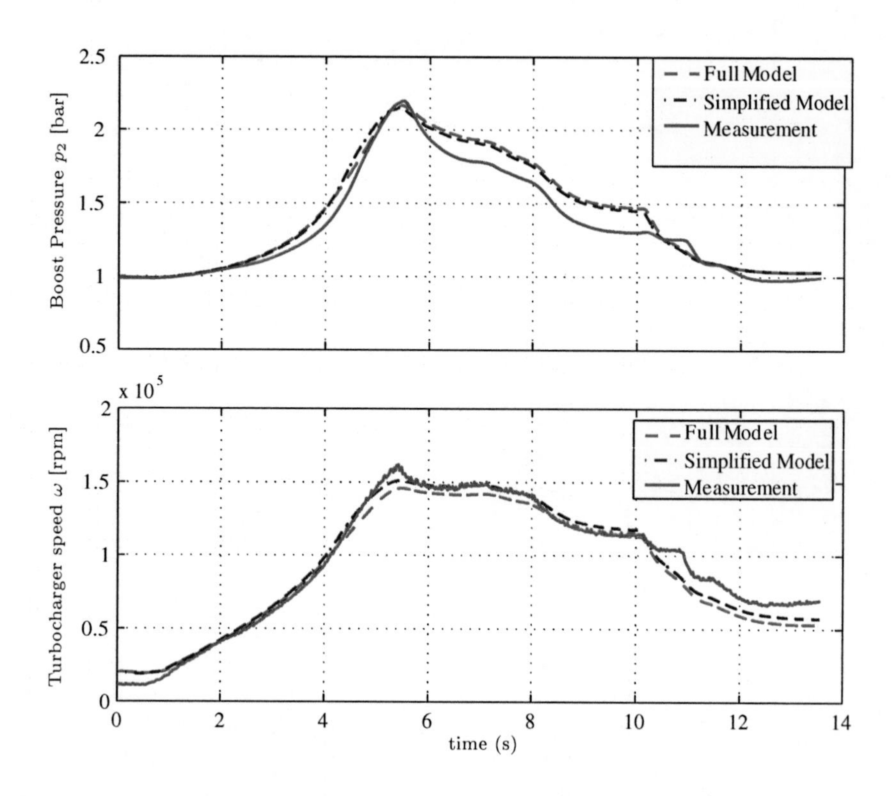

Fig. A.14: Accuracy of the simplified model of a one-stage turbocharged air-system.

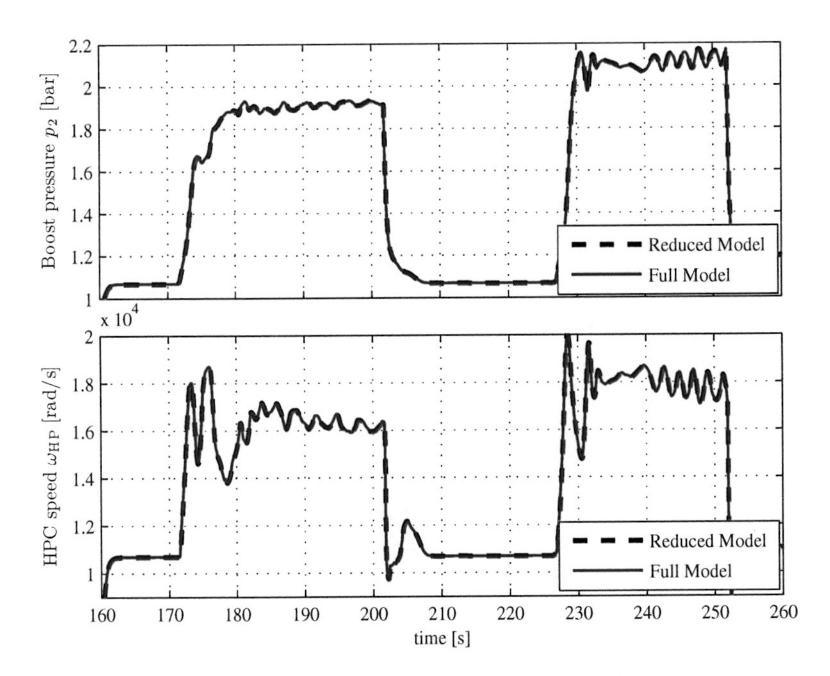

Fig. A.15: Accuracy of the simplified model of a two-stage turbocharged air-system.

B. OPEN PROBLEMS

B.1 Two-Degrees-of-Freedom IMC

B.1.1 Linear Case

A two-degrees-of-freedom control structure may improve results if both disturbance rejection and command tracking is desired. Therefore, a two degrees-of-freedom IMC control structure is reviewed and its design is briefly discussed. Note that this work mainly focuses on a one-degree-of-freedom design. The reader is referred to [9, 72] for more detail on two-degrees-of-freedom IMC design.

A two-degrees-of-freedom IMC structure is shown in Fig. B.1. The

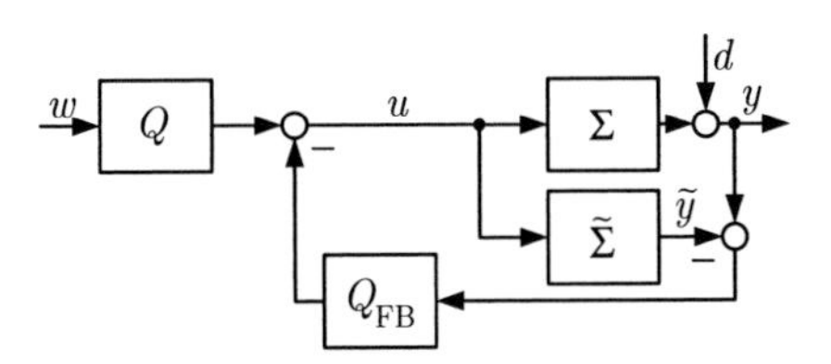

Fig. B.1: Two-degrees-of-freedom IMC structure.

control error is defined as $e = w - y$. From Fig. B.1 one gets

$$
\begin{aligned}
e(s) = {} & \frac{I - \widetilde{\Sigma}(s) Q_{\mathrm{FB}}(s)}{I + Q_{\mathrm{FB}}(s)\left(\Sigma(s) - \widetilde{\Sigma}(s)\right)} d(s) \\
& - \left[I - \frac{\Sigma(s) Q(s)}{1 + Q_{\mathrm{FB}}(s)\left(\Sigma(s) - \widetilde{\Sigma}(s)\right)} \right] w(s).
\end{aligned}
\tag{B.1}
$$

Considering an exact model ($\widetilde{\Sigma}(s) = \Sigma(s)$), one finds from Eq. (B.1) the relationship

$$e(s) = \Big(I - \Sigma(s)Q_{\mathrm{FB}}(s)\Big)d(s) - \Big(I - \Sigma(s)Q(s)\Big)w(s). \tag{B.2}$$

From the two equations above, one concludes that $Q(s)$ is to be designed for reference tracking and the feedback controller $Q_{\mathrm{FB}}(s)$ should be designed for disturbance rejection. Moreover, both can be designed independently of each other in the presence of small modelling errors.

Design. It is proposed to design $Q(s)$ as discussed in Section 2.2 with the exception that the designer does not need to consider a gain restriction on measurement noise. Further, $Q_{\mathrm{FB}}(s)$ must have the property $Q_{\mathrm{FB}}(0) = 1$ for the closed-IMC loop to guarantee zero steady-state offset. It is proposed to design $Q_{\mathrm{FB}}(s)$ as a low-pass filter

$$Q_{\mathrm{FB}}(s) = \frac{1}{(s/\lambda_{\mathrm{FB}} + 1)^k}, \tag{B.3}$$

where both k and λ_{FB} are to be chosen by the designer. The following rule of thumb applies: A slow feedback part $Q_{\mathrm{FB}}(s)$ (i. e., high k and small λ_{FB}) results in greater closed-loop robustness but slower command tracking in the presence of modelling errors and disturbances. A fast feedback part (i. e., small k and high λ_{FB}) results in aggressive disturbance rejection and good command tracking, but also increases noise amplification and the risk of instability.

Remark B.1. Assume a two-degrees-of-freedom IMC, where $Q(s)$ and $Q_{\mathrm{FB}}(s)$ are chosen to be identical (i. e., $Q(s) = Q_{\mathrm{FB}}(s)$). Then the resulting two-degrees-of-freedom IMC closed-loop behaviour is *identical* to the one-degree-of-freedom IMC closed-loop behaviour (as shown in Fig. 2.1) with IMC controller $Q(s)$.

B.1.2 Nonlinear Case

For the linear IMC, a two-degrees-of-freedom structure has been reviewed in Section B.1.1. For nonlinear IMC, it is proposed to exploit the separate implementation of IMC Filter F and right inverse $\widetilde{\Sigma}^{\mathrm{r}}$ (cf. Fig. 3.6).

Figure B.2 shows the proposed two-degrees-of-freedom IMC structure for nonlinear systems. The structure in Fig. B.2 is similar to the two-degrees-of-freedom structure for linear IMC in Fig. B.1. The two differ in

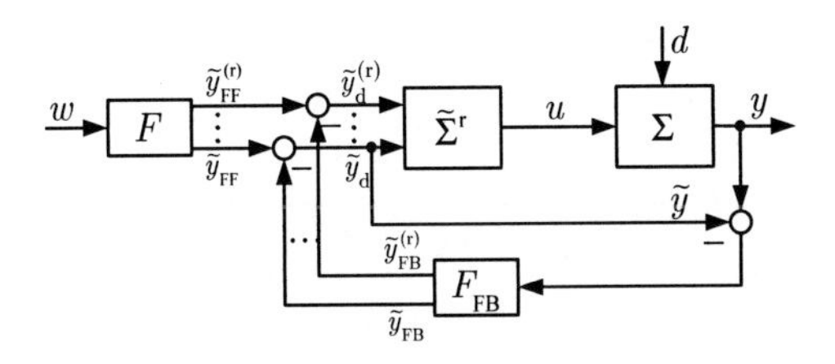

Fig. B.2: Two-degrees-of-freedom structure for a nonlinear IMC.

the location of the model inverse $\widetilde{\Sigma}^{\mathrm{r}}$. Due to the nonlinearity of the right inverse $\widetilde{\Sigma}^{\mathrm{r}}$, a summation of control inputs as presented in Fig. B.1 is not a feasible step since the principle of superposition does not hold in this case.

With the implementation in Fig. B.2, it is only necessary to implement the right inverse once. The blocks F and F_{FB} both represent state-variable-filters (cf. Fig. 3.7). However, their parameterisation may differ from each other. If both filters are chosen to be identical ($F = F_{\mathrm{FB}}$) then the two-degrees-of-freedom structure in Fig. B.2 behaves exactly as the one-degree-of-freedom IMC structure from Fig. 3.8.

Assuming an exact model $\widetilde{\Sigma} = \Sigma$ and assuming the presence of output disturbances (i. e., $y = \tilde{y} + d$) one finds

$$y = Fw + (I - F_{\mathrm{FB}})d. \tag{B.4}$$

Hence, the filter F can be designed for feedforward control (as in one-degree-of-freedom) and F_{FB} is to be designed for disturbance rejection. Both filters need to be of order r and need to be implemented as SVF. Note that the two filters can be designed independently of each other.

Since this work focuses on the one-degree-of-freedom structure, the two-degrees-of-freedom design is not discussed further.

B.2 Ad Hoc Non-Minimum Phase IMC Design for Flat Systems using a Minimum Phase Model

This section presents a ad hoc method by which NMP systems can be controlled using IMC. It is regarded as a "hands-on" (ad-hoc) solution due to the lack of a generally applicable algorithm.

Main idea. It is proposed to deal with NMP nonlinear systems in the same manner as it was introduced for linear systems in Section 2.2.2. Hence, it is proposed to do the following:

- Invert only a part $\widetilde{\Sigma}_{\mathrm{MP}}$ of the model $\widetilde{\Sigma}$ which leads to a stable inverse. Hence, the part $\widetilde{\Sigma}_{\mathrm{MP}}$ is MP.
- The steady-state gain of the part $\widetilde{\Sigma}_{\mathrm{MP}}$ to be inverted must be equal to the steady-state gain of the NMP model $\widetilde{\Sigma}$, i. e., $g(\widetilde{\Sigma}_{\mathrm{ss}}) = g(\widetilde{\Sigma}_{\mathrm{MP,ss}})$.
- In the IMC structure, the internal model $\widetilde{\Sigma}$ should *not* be replaced by a direct connection of the filter output $\tilde{y}_{\mathrm{d}}$.

It is proposed to obtain the IMC controller Q by

$$\boxed{Q = \widetilde{\Sigma}^{\mathrm{r}}_{\mathrm{MP}}\, F}, \tag{B.5}$$

where the IMC filter F is designed as proposed in Section 3.5.1. Figure

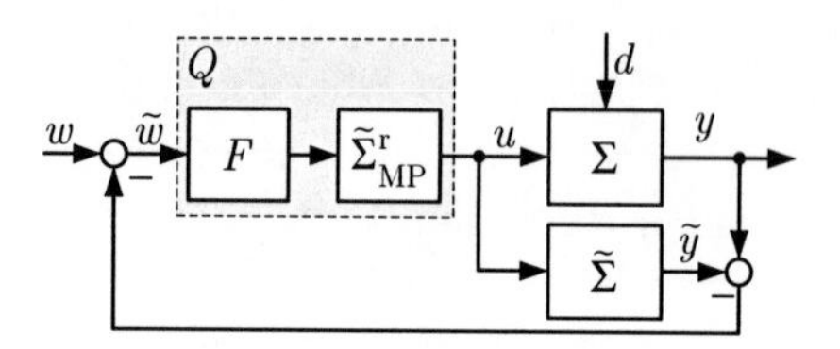

Fig. B.3: IMC structure for NMP nonlinear systems where the right inverse $\widetilde{\Sigma}^{\mathrm{r}}_{\mathrm{MP}}$ has been derived from a minimum-phase approximation of the non-minimum phase model $\widetilde{\Sigma}$.

B.3 shows the proposed IMC structure for the control of NMP nonlinear systems.

Interpretation. The altered model $\widetilde{\Sigma}_{\mathrm{MP}}$ needs to be chosen by the designer. The result is an inverse $\widetilde{\Sigma}^{\mathrm{r}}_{\mathrm{MP}}$ that is stable, but

$$\widetilde{\Sigma}\, \widetilde{\Sigma}^{\mathrm{r}}_{\mathrm{MP}} \neq I$$

holds. That is, the MP inverse $\widetilde{\Sigma}^{\mathrm{r}}_{\mathrm{MP}}$ is not a perfect inverse of the (full) NMP model $\widetilde{\Sigma}$. As in the linear case, this approach does not change the properties of nominal stability (Property 3.1) and steady-state offset

(Property 3.3). Naturally, the IMC design in Eq. (B.5) does influence the property of robust stability (Property 3.4), which is to be expected, since this is also true for linear systems. As in the linear case, the closed loop has an NMP I/O behaviour (cf. Eq. (2.29)). In this case, however, the NMP I/O behaviour may be nonlinear.

Unfortunately, the proposed procedure is difficult to follow, since no general algorithm can be given for the removal of the NMP behaviour: In the nonlinear case, a system representation as a transfer function does not exist. Thus, removal of the NMP behaviour of a nonlinear model is non-trivial.

Application to flat NMP models. A flatness-based IMC (see Theorem 4.1) for an NMP system yields an unstable IMC controller because the solution, obtained from the differential equation (4.10b) (called $F_{\mathrm{y}\to\mathrm{z}}$), is unstable. Thus, the following method is proposed:

> If the solution of $F_{\mathrm{y}\to\mathrm{z}}$ is unstable, one should *change* $F_{\mathrm{y}\to\mathrm{z}}$ such that
>
> 1. the altered differential equation $F_{\mathrm{y}\to\mathrm{z,MP}}$ yields a stable solution,
> 2. the steady-state gain of $F_{\mathrm{y}\to\mathrm{z}}$ is retained (i. e., $F_{\mathrm{y}\to\mathrm{z,ss}} = F_{\mathrm{y}\to\mathrm{z,MP,ss}}$), and
> 3. the order of $F_{\mathrm{y}\to\mathrm{z}}$ is retained.

The resulting inverse $\widetilde{\Sigma}^{\mathrm{r}}_{\mathrm{MP}}$ is stable and has the same steady-state gain as $\widetilde{\Sigma}$. Thus, the properties of nominal stability and zero steady-state offset of the IMC structure still hold.

Although it seems unnatural to directly alter the right inverse of a model, it is admissible since it is equivalent to inverting a different (minimum phase) model. Unfortunately, the method of *how* the operator $F_{\mathrm{y}\to\mathrm{z}}$ is changed to obtain a stable $F_{\mathrm{y}\to\mathrm{z,MP}}$ is completely left to the engineer. Deriving a generally applicable algorithm to do this task is an unsolved problem.

The following presents an example of this method to flat systems. In the example, a linear system is discussed. Nevertheless, the steps proposed for nonlinear systems are followed to demonstrate their feasibility and to give the reader a means for comparison to the linear case. A nonlinear example can be obtained by altering the following model (B.6) accordingly and then proceed by taking the same steps.

Example B.1 (Flatness-based IMC of an NMP system):
Consider the linear plant model

$$\dot{\boldsymbol{x}} = \begin{bmatrix} 0 & 1 \\ -5 & -2 \end{bmatrix} \boldsymbol{x} + \begin{bmatrix} 0 \\ 1 \end{bmatrix} u \tag{B.6a}$$

$$\tilde{y} = \begin{bmatrix} 5 & -5 \end{bmatrix} \boldsymbol{x} \tag{B.6b}$$

with poles at $p_{1/2} = -1 \pm 2j$ and a RHP zero at $z = +1$. Thus, the model is stable and NMP.

With the flat output $z = x_1$, one finds the control law

$$\psi_{\mathrm{u}}: \quad u = \ddot{z}_{\mathrm{d}} + 2\dot{z}_{\mathrm{d}} + 5z_{\mathrm{d}}. \tag{B.7}$$

Further, one finds the transformation from $\tilde{y}_{\mathrm{d}}$ to z_{d} as

$$\begin{aligned} F_{\mathrm{y}\to\mathrm{z}}: \quad & \tilde{y}_{\mathrm{d}} = 5z_{\mathrm{d}} - 5\dot{z}_{\mathrm{d}} \\ \Leftrightarrow \quad & \dot{\tilde{y}}_{\mathrm{d}} = 5\dot{z}_{\mathrm{d}} - 5\ddot{z}_{\mathrm{d}}. \end{aligned} \tag{B.8}$$

As expected, $F_{\mathrm{y}\to\mathrm{z}}$ is unstable due to a pole at $+1$ which is the location of the model zero. Thus, Eq. (B.8) is changed to

$$\begin{aligned} F_{\mathrm{y}\to\mathrm{z,MP}}: \quad & \tilde{y}_{\mathrm{d}} = 5z_{\mathrm{d}} + 5\dot{z}_{\mathrm{d}} \\ \Leftrightarrow \quad & \dot{\tilde{y}}_{\mathrm{d}} = 5\dot{z}_{\mathrm{d}} + 5\ddot{z}_{\mathrm{d}}, \end{aligned} \tag{B.9}$$

which yields a stable pole at -1 and is equivalent to inverting a different model, which had a zero at -1 to begin with.

The IMC filter is chosen as $F(s) = \frac{1}{s/\lambda+1}$ since the model's relative degree is $r = 1$. Its implementation as SVF is shown in Fig. B.4(a). The implementation of $F_{\mathrm{y}\to\mathrm{z,MP}}$ is shown in Fig. B.4(b). With the output of

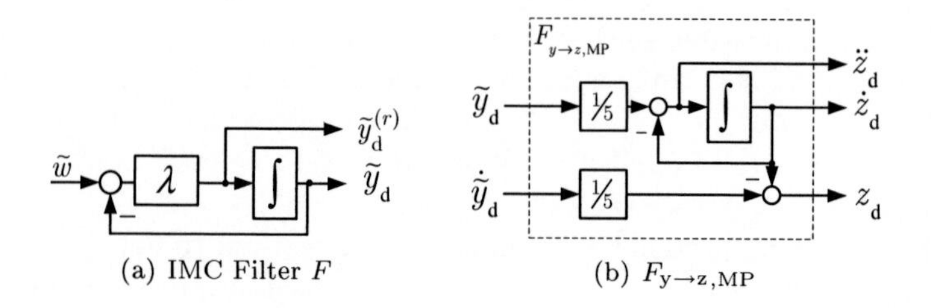

(a) IMC Filter F

(b) $F_{\mathrm{y}\to\mathrm{z,MP}}$

Fig. B.4: IMC filter F and $F_{\mathrm{y}\to\mathrm{z,MP}}$ for the example.

$F_{\mathrm{y}\to\mathrm{z,MP}}$, namely $z_{\mathrm{d}}, \dot{z}_{\mathrm{d}}, \ddot{z}_{\mathrm{d}}$, the input u can be computed with Eq. (B.7).

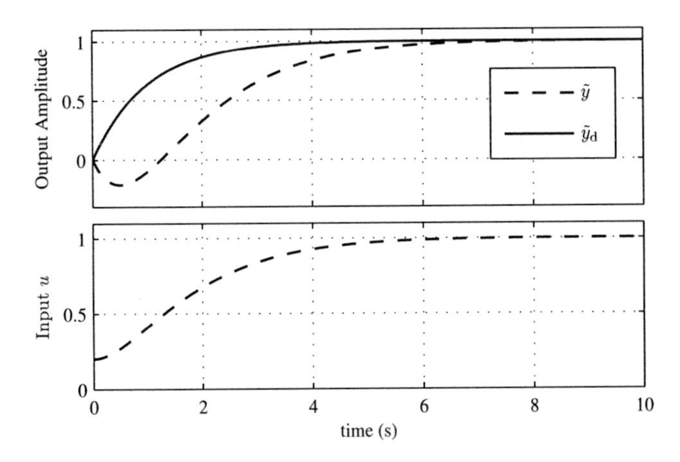

Fig. B.5: Closed-loop response of the IMC controller of Example B.1.

Finally, Fig. B.5 shows the response of the model output $\tilde{y}$ and the filter output $\tilde{y}_{\mathrm{d}}$. As expected, the two differ from each other in such that the model output $\tilde{y}$ still exhibits an inverse response which is characteristic for an NMP behaviour resulting from one right-half plane zero. However, they both reach the same steady-state value. As expected, the input u is a stable signal which indicates an internally stable feedforward controller Q. ■

BIBLIOGRAPHY

References Cited

[1] P. Apkarian and R. Adams. Advanced gain-scheduling techniques for uncertain systems. *IEEE Transactions on Control Systems Technology*, 6(1):21–32, 1998.

[2] R.D. Bartusiak, C. Georgakis, and M.J. Reilly. Nonlinear feedforward/feedback control structures designed by reference system synthesis. *Chemical Engineering Science*, 44(9):1837–1851, 1989.

[3] R. Bauder, S. Wild, A. Peters, L. Mikulic, H. Schulte, U. Mohr, and C. Morreale. Forum der Meinungen: Aufladung von Dieselmotoren in Pkw – Welche Systeme setzen sich durch? *Motorentechnische Zeitschrift (MTZ)*, 9:714–717, 2003.

[4] B.W. Bequette. Nonlinear control of chemical processes – a review. *Ind.Eng.Chem.Res.*, 30:1391–1413, 1991.

[5] B.W. Bequette. *Practical Approaches to Nonlinear control.* Nonlinear Model-based Process Control. Kluwer, Dordrecht, 1998.

[6] L. Bitauld, M. Fliess, and J. Lévine. A flatness based control synthesis of linear systems and application to windshield wipers. In *In Proc. 4th European Control Conference.*

[7] A. Bloch. Yellow-Express. *Auto-Motor-Sport*, 13:64–68, June 2006. Porsche Turbo testdrive.

[8] W. Bohl. *Technische Strömungslehre.* Vogel Buchverlag, 1994.

[9] C. Brosilow and B. Joseph. *Techniques of Model-Based Control.* Prentice Hall international series in the physical and chemical engineering sciences. Prentice Hall, Upper Saddle River, New Jersey, 2002.

[10] C.G. Cantemir. Twin turbo strategy operation. In *SAE 2001 World Congress*, pages 149–157, Detroit, 2001.

[11] R. Carbon. Seminar Automotive Software Engineering – Vorbesprechung. Fraunhofer IESE, 2006.

[12] D. Chen and B. Paden. Stable inversion of nonlinear non-minimum phase systems. *International Journal on Control*, 64(1):81–97, 1996.

[13] W.H. Chen. Analytic predictive controllers for nonlinear systems with ill-defined relative degree. In *IEE Proc.-Control Theory Appl.*, volume 148, pages 9–16, Januar 2001.

[14] G. Claeys, N. Retière, N. HadjSaid, P. Lemerle, E. Varret, and R. Belhomme. Dynamic modeling of turbo-charged diesel engine for power system studies. Technical Report 0-7803-6681-6/01, IEEE, 2001.

[15] A. Conn, N. Gould, and P. Toint. A globally convergent augmented Lagrangian algorithm for optimization. *SIAM Journal on Numerical Analysis*, 28(2):545–572, 1991.

[16] J. Conway. *Functions of One Complex Variable I.* Springer, 1986.

[17] C. Daniel and F. S. Wood. *Fitting Equations to Data.* John Wiley & Sons, New York, 1980.

[18] J. Daniels. BMW Renews its 6-cylinder Range. *Automotive Engineer*, pages 40–42, July/August 2004.

[19] P. Daoutidis and C. Kravaris. Structural evaluation of control configurations for multivariable nonlinear processes. *Chemical Engineering Science*, 47(5):1091–1107, 1992.

[20] P. Daoutidis and A. Kumar. Structural analysis and output feedback control of nonlinear multivariable processes. *AIChE*, 40(4):647–669, 1994.

[21] A. Datta. *Adaptive Internal Model Control.* Springer-Verlag, London, 1998.

[22] S. Devasia. Should model-based inverse inputs be used as feedforward under plant uncertainty? *IEEE Transactions on Automatic Control*, 47(11):1865–1871, 2002.

[23] S. Devasia, D.Chen, and B. Paden. Nonlinear inversion-based output tracking. *IEEE Transactions on Automatic Control*, 41(7):930–942, 1996.

[24] S. Devasia and B. Paden. Stable inversion for nonlinear nominimum-phase time-varying systems. *IEEE Transactions on Automatic Control*, 43(2):283–288, 1998.

[25] DieselNet Technology Guide. Diesel engine fundamentals. `http://www.dieselnet.com/tech/diesel_fund.html`, 2001.

[26] DieselNet Technology Guide. Turbochargers for diesel engines. `http://www.dieselnet.com/tech/diesel_turbo.html`, 2001.

[27] DieselNet Technology Guide. Advanced technologies: Air induction. `http://www.dieselnet.com/tech/engine_adv_air.html`, 2002.

[28] C.G. Economou and M. Morari. Internal model control. 5. extension to nonlinear systems. *Ind. Eng. Chem. Process Des. Dev.*, 25:403–411, 1986.

[29] M. Fliess, J. Lévine, P. Martin, and P. Rouchon. On differentially flat nonlinear systems. *Nonlinear Control Systems Design*, pages 408–412, 1992.

[30] M. Fliess, J. Lévine, P. Martin, and P. Rouchon. Flatness and defect of nonlinear systems: Introductory theory and examples. *International Journal of Control*, 61:1327–1361, 1995.

[31] P.M. Frank. *Entwurf von Regelkreisen mit vorgeschriebenem Verhalten.* Braun, Karlsruhe, 1974.

[32] C.E. Garcia and M. Morari. Internal model control. 1. A unifying review and some new results. *Ind. Eng. Chem. Process Des. Dev.*, 21:308–323, 1982.

[33] J. Garcia-Ortiz, P. Langthaler, and L. Del Re. GPC control of the airpath of high speed diesel engines. In *IEEE Conference on Control Applications*, 2006.

[34] D.N. Godbole and S.S. Sastry. Approximate decoupling and asymptotic tracking for mimo systems. *IEEE Transactions on Automatic Control*, 40(3):441–450, March 1995.

[35] K. Graichen and M. Zeitz. Inversionsbasierter Vorsteuerungsentwurf mit Ein- und Ausgangsbeschränkungen. *Automatisierungstechnik*, 4:187–199, 2006.

[36] Knut Graichen. *Feedforward Control Design for Finite-Time Transition Problems of Nonlinear Systems with Input and Output Constraints.* Shaker, 2006.

[37] GT Power. Gamma Technologies homepage. Internet. `http://www.gtisoft.com/`.

[38] G. Hack. Der Doppler-Effekt. *auto motor sport*, 4:60–61, 2004.

[39] V. Hagenmeyer. *Robust Nonlinear Tracking Control Based on Differential Flatness.* VDI Verlag, Düsseldorf, 2002.

[40] V. Hagenmeyer and M. Zeitz. Flachheitsbasierter Entwurf von linearen und nichtlinearen Vorsteuerungen. *Automatisierungstechnik*, 52:3–12, 2004.

[41] M.A. Henson. *Nonlinear Process Control.* Prentice Hall, 1997.

[42] M.A. Henson and D.E. Seborg. Input-output linearization of general nonlinear processes. *AICHE Journal*, 36(11):1753–1757, November 1990.

[43] M.A. Henson and D.E. Seborg. A critique of exact linearization strategies for process control. *J. Process Control*, 1:122–139, 1991.

[44] M.A. Henson and D.E. Seborg. An internal model control strategy for nonlinear systems. *AIChE Journal*, 37(7):1065–1081, 1991.

[45] M. Herceg, T. Raff, R. Findeisen, and F. Allgöwer. Nonlinear model predictive control of a turbocharged diesel engine (i). In *IEEE Conference on Control Applications*, 2006.

[46] H. Heuser. *Gewöhnliche Differentialgleichungen.* B.G. Teubner Stuttgart, 1995.

[47] J. Heywood. *Internal Combustion Engine Fundamentals.* McGraw-Hill, 1988.

[48] R.M. Hirschorn. Invertibility of multivariable nonlinear control systems. *IEEE Transactions on Automatic Control*, AC-24:855–865, 1979.

[49] R.M. Hirschorn. Output tracking through singularities. In *Proceedings of the 41st IEEE Conference on Decision and Control*, pages 3843–3848, 2002.

[50] J.M. Horowitz. *Synthesis of Feedback Systems.* Academic Press, 1963.

[51] A. Isidori. *Nonlinear Control Systems.* Springer, New York, 3rd edition, 1995.

[52] M. Jung. *Mean-Value Modelling and Robust Control of the Airpath of a Turbocharged Diesel Engine.* PhD thesis, University of Cambridge, 2003.

[53] D.C. Karnopp, D.L. Margolis, and R.C. Rosenberg. *System Dynamics.* John Wiley & Sons, 4th edition, 2006.

[54] H.K. Khalil. *Nonlinear Systems.* Prentice-Hall, New Jersey, 3rd edition, 2000.

[55] U. Kiencke and L. Nielsen. *Automotive Control Systems.* Springer, 2nd edition, 2005.

[56] K. Knopp. *Theory of Functions, Parts I and II (Dover Books on Mathematics).* Dover Publications, 1996.

[57] P. Kokotovic and M. Arcak. Constructive nonlinear control: a historical perspective. *Automatica*, 37:637–662, 2001.

[58] C. Kravaris and Y. Arkun. Geometric nonlinear control - an overview. In *Chemical Process Control*, volume 4, pages 477–515, 1991.

[59] C. Kravaris and P. Daoutidis. Nonlinear state feedback control of second-order nonmiminum-phase nonlinear systems. *Computers Chem. Engng*, 14(4/5):439–449, 1990.

[60] C. Kraviris and J.C. Kantor. Geometric methods for nonlinear process control. 1. background. *Ind. Eng. Chem. Res.*, 29:2295–2310, 1990.

[61] C. Kraviris and J.C. Kantor. Geometric methods for nonlinear process control. 2. controller synthesis. *Ind. Eng. Chem. Res.*, 29:2310–2323, 1990.

[62] D.J. Leith and W.E. Leithead. Input-output linearisation of nonlinear systems with ill-defined relative degree: The ball & beam revisited. Report, Dept. Electronics & Electrical Engineering, University of Strathclyde, Glasgow, 2001.

[63] J. Lunze. *Regelungstechnik 1.* Springer, Berlin, 2004.

[64] J. Lunze. *Regelungstechnik 2.* Springer, Berlin, 2005.

[65] L. Marconi and A. Isidori. Mixed internal model-based and feedforward control for robust tracking in nonlinear systems. *Automatica*, 36:993–1000, 2000.

[66] D. Marquardt. An algorithm for least squares estimation of nonlinear parameters. *SIAM J. Appl. Math.*, 11:431–441, 1963.

[67] F. Mayinger and K. Stephan. *Thermodynamik 1 - Einstoffsysteme.* Springer, Berlin, 1992.

[68] W.J. Mccolm and M.T. Tham. On globally linearising control about singular points. In *Proceedings of the American Control Conference*, pages 2229–2233, June 1995.

[69] P. McLellan. A differential-algebraic perspective on nonlinear controller design methodologies. *Chemical Engineering Science*, 49(10):1663–1679, 1994.

[70] P. Moraal and I. Kolmanovsky. Turbocharger modeling for automotive control applications. Number SAE 1999-01-0908, 1999.

[71] P. Moraal, Van M.J. Nieuwstadt, and I.V. Kolmanovsky. Model based control of diesel engines. In *Tagungsband, 2. Symposium: Steuerungssysteme für den Antriebsstrang von Kraftfahrzeugen*, pages 85–95. Berlin, 1999.

[72] M. Morari and E. Zafiriou. *Robust Process Control.* Prentice Hall, Englewood Cliffs, New Jersey, 1989.

[73] G.C. Newton, L.A. Gould, and J.F. Kaiser. *Analytic Design of Feedback Controls.* John Wiley & Sons, 1957.

[74] M.I. Nieuwstadt, I.V. Kolmanovsky, P. Moraal, A. Stefanopoulou, and Jankovic. EGR-VTG control schemes: Experimental comparison for a high-speed diesel engine. *IEEE Control Systems Magazine*, June:64–79, 2000.

[75] M.I. Nieuwstadt and R.M. Murray. Real-time trajectory generation for differentially flat systems. *International Journal of Robust and Nonlinear Control*, 8(11):995–1020, 1998.

[76] R. Nitsche. Zwischenbericht zum DS-Projektteil FLI114: Modellgestützte Regelung von Luftsystemen. Internal Report 2190, Robert Bosch GmbH, 2002.

[77] J. Nocedal and S. Wright. *Numerical Optimization. Springer Series in Operations Research.* Springer, 2nd edition, 2006.

[78] P. Ortner, P. Langthaler, J. Garcia-Ortiz, and L. del Re. MPC for a diesel engine airpath using an explicit approach for constraint systems. In *IEEE Conference on Control Applications*, 2006.

[79] H. Paffrath. *Untersuchungen zum Potential eines regelbaren zweistufigen Aufladeverfahrens für Nutzfahrzeugmotoren.* PhD thesis, RWTH Aachen, 1995.

[80] W. Respondek. *Nonlinear Controllability and Optimal Control*, chapter Right and left invertibility of nonlinear control systems, pages 133–176. Marcel Dekker Inc., New York and Basel, 1990.

[81] R. Rothfuß. *Anwendung der flachheitsbasierten Analyse und Regelung nichtlinearer Mehrgrößensysteme.* Number 664 in Reihe 8. VDI Verlag GmbH, Düsseldorf, 1997.

[82] A. Schloßer. *Modellbildung und Simulation zur Ladedruck- und Abgasrückführregelung an einem Dieselmotor.* PhD thesis, RWTH Aachen, Aachen, 2000.

[83] R. Sepulchre. *Constructive Nonlinear Control.* Springer, 1997.

[84] M.M. Seron, S.F. Graebe, and G.C. Goodwin. All stabilizing controllers, feedback linearization and anti-windup: A unified review. In *Proceedings of the American Control Conference*, 1994.

[85] J.S. Shamma and M. Athans. Guaranteed properties of gain scheduled control for linear parameter-varying plants. *Automatica*, 27(3):559–564, 1991.

[86] G. Silva, A. Datta, and S.P. Bhattacharyya. On the Stability and Controller Robustness of Some Popular PID Tuning Rules. *IEEE Transactions on Automatic Control*, 48(9):1638–1641, 2003.

[87] S. Skogestad and I. Postlethwaite. *Multivariable feedback control.* John Wiley & Sons, New York, 2004.

[88] O.J.M. Smith. Closer control of loops with dead time. *Chem. Eng. Progress*, 53(5), 1975.

[89] Star CD. STAR-CD homepage. Internet. `http://www.cd-adapco.com/`.

[90] F. Steinparzer, W. Stütz, H. Kratochwill, and W. Mattes. Der neue BMW-Sechszylinder-Dieselmotor mit Stufenaufladung. *Motorentechnische Zeitschrift (MTZ)*, 5:334–344, May 2005.

[91] K. Takaba. Analysis and synthesis of anti-windup control system based on Youla parametrization. In *Proceedings of the 41st IEEE Conference on Decision and Control*, 2002.

[92] G. Throtokatos and N.P. Kyrtatos. Diesel engine transient operation with turbocharger compressor surging. Technical Report 2001-01-1241, SAE, 2001.

[93] J. Tsianias and N. Kalouptsidis. Invertibility of nonlinear analytic single-input systems. *IEEE Transactions on Automatic Control*, AC-28(9):931–933, 1983.

[94] J. von Löwis and J. Rudolph. Real-time trajectory generation for flat systems with constraints. *Nonlinear and Adaptive Control, LNCIS*, (281):385–394, 2003.

[95] M. Wellers and M. Elicker. Regelung der AGR-Rate und des Ladedruckes mit Hilfe eines nichtlinearen Modellbasierten prädiktiven Reglers. In Rolf Isermann, editor, *Steuerung und Regelung von Fahrzeugen und Motoren – AUTOREG*, pages 99–108, Düsseldorf, 2004. VDI Verlag GmbH.

[96] A. Wittmer, P. Albrecht, B. Becker, G. Vogt, and R. Fischer. *Zweistufige Aufladung eines Pkw-Dieselmotors*, chapter 1, pages 1–10. expert verlag, 2004.

[97] P.C. Young. Parameter estimation for continuous-time models – a survey. *Automatica*, 17:23–39, 1981.

[98] R. Yui. *Continuous-Time Model Reference Controller Design.* PhD thesis, New Mexico State University, 1995.

[99] G. Zames. On the input-output stability of time-varying nonlinear feedback systems part i and ii. *IEEE Transactions on Automatic Control*, AC-11(2 and 3):228–238 and 465–476, 1966.

[100] R. Zanasi, C. G. Bianco, and A. Tonielli. Nonlinear filters for the generation of smooth trajectories. *Automatica*, 36:439–448, 2000.

[101] S. Zazueta and J. Alvarez. A robust internal model controller for nonlinear systems. In *Proceedings of the American Control Conference*, pages 561–566, 2002.

[102] M. Zeitz, K. Graichen, and T. Meurer. Vorsteuerung mit Trajektorienplanung als Basis einer Folgeregelung. In *GMA-Kongress "Automation als interdisziplinäre Herausforderung"*, June 2005.

[103] M. Zeitz, R. Rothfuß, and J. Rudolph. Flachheit: Ein neuer Zugang zur Analyse und Regelung nichtlinearer Systeme. *Automatisierungstechnik*, (54):517–525, 1997.

[104] A. Zheng, M.V. Kothare, and M. Morari. Anti-windup design for internal model control. *International Journal of Control*, 60(5):1015–1024, 1994.

[105] K.A. Zinner. *Aufladung von Verbrennungsmotoren*. Springer-Verlag, 1985.

Contributions Generated

Conferences.

[i] J. Hanschke, R. Nitsche, and D. Schwarzmann. Nonlinear internal model control of diesel air systems. In *E-COSM – Rencontres Scientifiques de l'IFP*, pages 121–131. Institut Francais du Petrole, October 2006.

[ii] D. Schwarzmann, J. Lunze, and R. Nitsche. A flatness-based approach to internal model control of linear systems. In *Proceedings of the American Control Conference*, 2006.

[iii] D. Schwarzmann, R. Nitsche, and J. Lunze. Diesel boost-pressure control using flatness-based internal model control. *SAE Special Publication Papers*, SP-2003(2006-01-0855), April 2006. Presented at SAE-World Conference 2006.

[iv] D. Schwarzmann, R. Nitsche, and J. Lunze. Modelling of the airsystem of a two-stage turbocharged diesel engine. In *MATHMOD*, 2006.

[v] D. Schwarzmann, R. Nitsche, J. Lunze, and A. Schanz. Pressure control of a two-stage turbocharged diesel engine using a novel nonlinear imc approach. In *Proceedings of the Conference on Control Applications*, 2006.

[vi] D. Schwarzmann, R. Nitsche, J. Lunze, and M. Schmidt. Nonlinear multivariable robust internal model control of a two-stage turbocharged diesel engine. In *Fifth IFAC Symposium on Advances in Automotive Control*, 2007.

[vii] D. Schwarzmann, Nitsche R., and J. Lunze. Robuste Ladedruckregelung eines Pkw-Dieselmotors mittels flachheitsbasierter IMC-Regelung. In *VDI AUTOREG*, 2006.

[viii] R. Nitsche and D. Schwarzmann. Flachheitsbasierte IMC-Regelung einer elektronisch kommutierten Synchronmaschine. In *Mechatronik*, May 2007.

Journals.

[ix] D. Schwarzmann, R. Nitsche, and J. Hanschke. Nonlinear internal model control of diesel air systems. *Oil & Gas Science and Technology-Revue de l'Institut Francais du Petrole (OGST)*, 63, 2007.

Internal Reports and Patents.

[x] D. Schwarzmann. Modellbasierte Funktionsentwicklung am Beispiel eines Pkw-Luftsystems. Internal report, Robert Bosch GmbH, Juni 2004.

[xi] D. Schwarzmann. *(Nonlinear) Internal-Model-Control: Untersuchung bestehender Konzepte und Einführung einer neuen Idee.* Internal report, Robert Bosch GmbH, März 2005.

[xii] D. Schwarzmann. Verfahren zum Betreiben einer Brennkraftmaschine, Computerprogramm-Produkt, Computerprogramm und Steuer- und/oder Regeleinrichtung. Patent 0710028.5-1263, Robert Bosch GmbH, 2007.

[xiii] D. Schwarzmann. Extrapolation of trubine mass flow. Applied for Patent, Robert Bosch GmbH, 2006.

Student Theses and Internships

[xiv] S. Hesse. Modellierung und Identifikation von Lusftsystemen turboaufgeladener Dieselmotoren. Diploma Thesis, Universität Stuttgart, July 2004.

[xv] R. Huck. Analyse eines turboaufgeladenen Diesel-Luftsystems und methodischer Entwurf einer robusten Ladedruckregelung. Diploma Thesis, Universität Stuttgart, May 2005.

[xvi] A. Schanz. Entwurf einer Vorsteuerung für das zweistufig aufgeladene Luftsystem. Diploma Thesis, Universität Stuttgart, October 2005.

[xvii] M. Schmidt. Erstellung und Durchführung eines modellbasierten Gesamtkonzeptes zur Identifikation, Funktionsentwicklung und Implementierung für die zweistufige Aufladung. Diploma Thesis, Universität Stuttgart, Januar 2006.

[xviii] B. Ahrens. Steer-by-Wire für Landmaschinen bis 60km/h. Diploma Thesis, Universität Karlsruhe, March 2007.

[xix] D. Daumiller. Positionsregelung eines hydraulischen arbeitsarms mittels nichtlinearem IMC. Diploma Thesis, Universität Stuttgart, June 2007.

[xx] K. Boualem. Extrapolation of compressor-data. Internschip, Robert Bosch GmbH, August 2004.

[xxi] T. Schlage. Modellierung eines Turboladers. Internship, Robert Bosch GmbH, August 2005.

[xxii] R. Huck. Automatisierte Applikation der Ladedruckregelung eines Diesel-Luftsystems. Internship, Robert Bosch GmbH, August 2004.

[xxiii] F. Kroll. Werkzeuge zur Modellierung von Turbinen und Kompressoren auf Basis von Herstellerdaten. Internship, Robert Bosch GmbH, Januar 2006.

[xxiv] J. Hanschke. Flachheitsbasierte IMC-Regelung von Diesel Luftsystemen mit Stellgrößenbeschränkungen. Internship, Robert Bosch GmbH, July 2006.

Dieter Schwarzmann
dieter.schwarzmann@gmail.com
+49 172 6241420

Home Address:

Eckenerstr. 63
74081 Heilbronn
Germany

Personal

Nationality: German

Education

7/2003 – Present	Ruhr-University, Bochum, Germany Ph.D. Student • Dissertation: "Nonlinear Internal Model Control with Automotive Applications"
8/2001 – 1/2003	Rose-Hulman Institute of Technology, Terre Haute, IN, USA • Master of Science in Mechanical Engineering, • Title of Masters' Thesis: "Optimal Pulse-Jet Control of an Atmospheric Rocket"
10/1997 – 5/2003	Universität Stuttgart, Germany • Dipl.-Ing. Engineering Cybernetics

---

Work Experience

7/2003 – Present	R&D Control Engineer Robert Bosch GmbH, Stuttgart, Germany • Modeling of turbocharged engines using Simulink and C++ • Control design **Since 06.2006:** Project leader "Control of mobile hydraulic robots"
2/2003 – 6/2003	Intern Robert Bosch GmbH, Stuttgart, Germany • Design and implementation of a positional controller for an electric throttle

---

Skills

Languages	German:	Native Speaker
	English:	Fluent
	French:	Intermediate
Technical	C/C++, Visual C++, Matlab/Simulink, Mathematica, AMESim, Maple, Ansys, EES, Turbo Pascal, Oberon, SQL, PHP